"十三五"普通高等教育本科规划教材

输电线路工程系列教材

输电线路工程专业英语

主编 唐 波 盛春来
编写 张宇娇 胡 刚 黄 力
　　　 智 李 高 亮 南春雷
主审 王璋奇

内 容 提 要

本书为"十三五"普通高等教育本科规划教材。本书是为满足高等院校输电线路专业（方向）的专业英语教学需求而编写的专业英语教材。全书共分为 10 个单元，每个单元侧重于一个主题；同时，全书也可分为 2 大部分，第一部分为精读，以输电线路专业知识体系的教学为主，共 7 个单元，分别为输电线路概论、架空导线力学、电晕放电及影响、输电线路工程设计方法、输电线路工程施工技术、输电线路典型缺陷及其处理、输电线路的雷击与保护；第二部分为泛读，以当前我国输电行业的一些发展概况为主，共 3 个单元，分别为电力系统的安全稳定、中国智能电网的发展，以及国家电网公司和南方电网公司的简介，供有兴趣的教师、技术人员和学生自学。

本书主要作为高等院校输电线路专业（方向）的教材，也可作为科技人员和工程技术人员学习专业英语的参考用书。

图书在版编目（CIP）数据

输电线路工程专业英语/唐波，盛春来主编. —北京：中国电力出版社，2017.9
"十三五"普通高等教育本科规划教材. 输电线路工程系列教材
ISBN 978-7-5198-0750-4

Ⅰ. ①输… Ⅱ. ①唐…②盛… Ⅲ. ①输电线路－电力工程－英语－高等学校－教材 Ⅳ. ①TM726

中国版本图书馆 CIP 数据核字（2017）第 200270 号

出版发行：中国电力出版社
地　　址：北京市东城区北京站西街 19 号（邮政编码 100005）
网　　址：http://www.cepp.sgcc.com.cn
责任编辑：牛梦洁（010-63412528）贾丹丹
责任校对：朱丽芳
装帧设计：赵姗姗
责任印制：吴　迪

印　　刷：北京雁林吉兆印刷有限公司
版　　次：2017 年 9 月第一版
印　　次：2017 年 9 月北京第一次印刷
开　　本：787 毫米×1092 毫米　16 开本
印　　张：13.75
字　　数：333 千字
定　　价：33.00 元

版 权 专 有　侵 权 必 究

本书如有印装质量问题，我社发行部负责退换

前　　言

输电线路专业伴随着我国电网建设的发展而发展。在 20 世纪 80 年代末，原电力部根据电力企业的需求，在当时的葛洲坝水电工程学院（现三峡大学）和东北电力学院（现东北电力大学）设置了输电线路工程全日制专科，同时在多所职工大学设置输电线路专业进行在职培训。时至今日，虽然输电线路专业未纳入教育部目录，但在输电行业持续旺盛的用人需求背景下，依靠诸多输电线路专业教育界同行的不懈努力，输电线路专业（方向）的高等教育研究与实践已经形成完整的硕士、本科及专科系列教育体系，凝练并形成了较完善的专业知识架构，为我国电网建设培养了大量高级工程技术人才。然而遗憾的是，在全国输电线路专业专业教材体系中尚无输电线路专业的专业英语教材。这导致各高校在专业英语教学实践中不得不引用其他专业的专业英语教材。现行的这些专业英语教材对输电线路的介绍寥寥，无法满足输电线路专业高级技术人才的专业英语学习需求。为此，编者基于多年输电线路专业的教学科研实践，萌发了编写本书的想法，并付诸行动，试图提供一本既能用于学生持续学习英语语言的表达能力和技巧，又可以介绍输电线路专业的英语词汇表达、国内外输电线路科学技术及现有输电行业发展的专业英语教材，从而让学生进一步了解输电专业，掌握常用科技英语的写作方法，并了解全英文科学研究论文的写作要求。

本书的编写依托于三峡大学对输电线路专业知识体系的凝练，即输电线路专业以力学、电学为理论基础，以输电线路工程的设计、施工和运行维护为专业核心技能。因此，结合对输电行业的一些介绍，全书可分为 2 个部分：第一部分为上述输电专业知识体系教学的精读内容，供教师重点教学和学生重点学习；第二部分为以输电行业及其发展介绍的泛读内容，供有兴趣的教师、技术人员和学生参考。

本书的特色如下：

（1）目标明确，即完全以输电线路相关专业知识的英语表达为编写对象，不但让学生借助专业知识进一步学习基础英语；同时，也让学生掌握本专业最新的科学技术发展状况，以及今后自学国外相关文献的技术手段。

（2）形式广泛，除常规专业英语中的课文、阅读和翻译形式外，在每个单元的后面还增加专业词汇、翻译技巧、英文科研论文撰写要求，以及相关的各种练习题。如此，让学生在边学边练的过程中，将英语技能和专业知识有机地结合在一起。

（3）教学层次分开，整本书分为精读和泛读两个部分，从而让教师教学和学生学习增加了选择余地，即学生可以仔细学习精读部分内容，自学泛读部分内容。

（4）增加了对科研论文的讲解，在学习专业英语的基础上，对科研论文的写作方法及要求进行了介绍，让学生了解科技论文写作和投稿的相关知识，更好地促进学生接触输电线路学科的科技前沿。

最后，编者想以诗经中的"嘤其鸣矣，求其友声"作为结语。本书应该是输电线路类专业的第一本专业英语教材，应该说能初步解决以往没有这类专业英语教材的问题。但由于能力水平有限，在此恳请各界同行对书中的内容提出修改建议，以期待后续更好地服务于高等院校输电线路专业英语的教学，以及科技人员与工程技术人员专业英语的学习。

本书由三峡大学唐波、盛春来担任主编，张宇娇、胡刚、黄力、智李、高亮和南春雷协助选稿并参与编写，盛春来对英文翻译进行整体审校。

本书由华北电力大学王璋奇教授主审。在此，对王璋奇教授的辛勤工作和支持表示特别感谢。

限于编者水平，书中难免有不妥和疏漏之处，恳请读者批评指正。

编　者

2017 年 7 月

Preface

Power transmission line develops with the advancement of electric network. At the end of the 1980s, the major of power transmission line of the full time technological college was established by the electric department in GeZhouBa College of Hydroelectric Engineering (now China Three Gorges University, CTGU) and Northeast Electric Power College (now Northeast Electric Power University) in accordance with the increasing needs of the power enterprises, and the major was also set up for the on-the-job training workers in many universities for employees. Till now, although the major of power transmission line has not registered in the catalogs of the Ministry of Education, it has developed a comprehensive system including postgraduate, undergraduate and technological academy education of this trade due to thriving demand for this major and the consistent efforts of the fellow teachers in this area, and it has formed a relatively perfect structure of this major, and cultivated a lot of advanced engineering technology workers for the power network construction in China. However, it is a pity that there is still no English course book for students majoring in power transmission line. Therefore, many students in this major have to use non-professional English course books in their professional English studies. In these course books, there is little introduction to power transmission line, which can not meet the professional need of their English studies for the advanced engineering technology workers in the field of power transmission line.

Therefore, based on years of teaching and researching in this major, the editors ventured to edit this book in want of providing a book for students who can consistently learn English expressions, skills and techniques, and learn English vocabularies and scientific techniques in this field at home and abroad. And this book has introduced the current development of power transmission line, enabling the students to learn further about this field and mastering the writing skills of science and technique English as well as the writing requirement of scientific papers in English.

This book is edited on the basis of the condensed professional knowledge by the teachers in CTGU. The theoretical bases are mechanics and electricity, and the core techniques are the design, construction and operation maintenance of power transmission line. So the book has two parts. Part one is an intensive introduction to the power transmission line for teachers and students to use in class. Part two is an extensive introduction to power transmission line and its development for the optional use of the teachers, students and people in need.

The characteristics of this book are as follows:

(1) Clearly targeted. It is totally edited in English for those in the power transmission line, which would facilitate the basic English studies with the professional knowledge, enabling the

students to grasp the newest scientific and technologic development in their profession, and to learn the skill of using foreign literatures.

(2) Extensive forms. Except for the ordinary texts, readings and translations in professional English books, there are professional vocabularles, translation skills, requirement of English scientific research papers and relevant exercises at the end of each chapter. So the students can combine their English techniques with professional knowledge in their studies.

(3) Separation of teaching arrangement. It contains the intensive and extensive parts, so both the teachers and students have their own choices. The intensive part is for in-class studies, and the extensive for after-class reading.

(4) Adding introduction to scientific research papers. In addition to learning professional English, this book has introduced the writing skills and requirements of scientific research papers, allowing them to learn about writing and contribution of scientific papers in a better acquaintance with the scientific frontier technology in power transmission line.

Finally, I would like to conclude with a sentence from the Book of Poetry in ancient China: the little bird is tweeting for a bosom friend. This book has primitively made an attempt in firstly editing a professional English course book in power transmission line. So there is inevitable to have some incorrect or imperfect expressions in the book owing to the limited knowledge of the editors, hence your valuable advice will be sincerely welcomed, and it will better serve for professional English teaching and learning in this major in Chinese universities as well as for the scientific and engineering workers in their professional English studies.

This book was chiefly edited by Tang Bo and Sheng Chunlai in CTGU. Zhang Yujiao, Hu Gang, Huang Li, Zhi Li, Gao Liang and Nan Chunlei also contributed to the edition. The English translation was examined and revised by Sheng Chunlai.

I am greatly indebted to Professor Wang Zhangqi, from North China Electric Power University, who examined and verified the edition.

The Editor
July, 2017

Contents

Preface

Chapter 1 Summary of Power Transmission Line ·· 1
    Section 1   Research on the Applicable Range of AC and DC
                 Transmission Voltage Class Sequence ····································· 1
    Section 2   An Overview of Power Transmission Systems in China ·············· 7
    Section 3   HVDC Transmission Systems ·· 14

Chapter 2 Mechanics of Overhead Power Transmission Line ························· 22
    Section 1   Introduction to Mechanics of Materials ································· 22
    Section 2   Introduction of the Structural Mechanics ······························ 27
    Section 3   Unbalanced Tension Analysis for UHV Transmission
                 Towers in Heavy Icing Areas ·· 33

Chapter 3 Corona Discharge ·· 40
    Section 1   Comparison of Methods for Determining Corona Inception
                 Voltages of Transmission Line Conductors ····························· 40
    Section 2   Air-Gap Discharge Characteristics in Foggy Conditions
                 Relevant to Lightning Shielding of Transmission Lines ············· 47
    Section 3   Short-circuit Current ·· 55

Chapter 4 The Design of Power Transmission Line ······································ 64
    Section 1   Overview of the Transmission Line Design Process ················· 64
    Section 2   Quantifying Siting Difficulty: A Case Study of US
                 Transmission Line Siting ··· 72
    Section 3   Reliability-Based Transmission Line Design ·························· 79

Chapter 5 Construction Technology ··· 86
    Section 1   Overview of the Transmission Line Construction Process ········· 86
    Section 2   Externality Identification and Quantification of
                 Transmission Construction Projects ······································ 93
    Section 3   Some Introductions about The Transmission Line Construction ······· 99

Chapter 6 Typical Defects from Power Transmission Line ····················· 106
    Section 1   An Overview of the Condition Monitoring of Overhead Lines ········ 106
    Section 2   Effect of Desert Environmental Conditions on The Flashover
                 Voltage of Insulators ·· 114
    Section 3   Failures in Outdoor Insulation Caused by Bird Excrement ········· 123

Chapter 7　Lightning Protection and Grounding ··· 130
    Section 1　Optimization of Hellenic Overhead High-voltage
               Transmission Lines Lightning Protection (1) ·································· 130
    Section 2　Optimization of Hellenic Overhead High-voltage
               Transmission Lines Lightning Protection (2) ·································· 138
    Section 3　Modeling of Power Transmission Lines for Lightning
               Back Flashover Analysis ··· 145

Chapter 8　Stability Problems of Power Grid ·· 153
    Section 1　Major Stability Problems of Long Distance and Large Capacity High
               Voltage AC/DC Transmission Systems and Interconnected Power Systems ······· 153
    Section 2　Technology of Improving Transmission Line Capacity and
               AC/DC Transmission System Stability Level ································· 161
    Section 3　Power Grids and Grid Interconnections in China ······························ 169

Chapter 9　The Development of China Smart Grid ·· 174
    Section 1　How Green Is The Smart Grid? ·· 174
    Section 2　Defining The Smart Grid ·· 178
    Section 3　The Expanded Scenario ·· 184

Chapter 10　The Introduction of SGCC and CSG ·· 191
    Section 1　Brief Introduction of State Grid Corporation of China (SGCC) ··············· 191
    Section 2　SGCC Profit and Output on The Rise ·· 198
    Section 3　China Southern Power Grid ··· 202

References ··· 209

Chapter 1 Summary of Power Transmission Line

Section 1 Research on the Applicable Range of AC and DC Transmission Voltage Class Sequence

The distribution of energy resources and energy demands is disharmonious in China. It objectively needs to optimize energy disposition on a large scale. Today, China has succeeded in the construction and operation of UHV AC and DC power transmission projects, therefore, the HV power transmission network mainly consists of the 1000 kV UHV and 500 kV EHV AC transmission voltage class sequence and ± 500, ± 660, ± 800, ± 1100 kV HVDC voltage class sequence. Since there are differences between AC and DC transmission in the aspects of project investment, environmental impact and reliability, it's significant and necessary to analyze and discuss the rational applicable range of each AC and DC transmission voltage grade from the view of economic and technical characteristics in the planning and construction of power grid.

For the applicable range of AC and DC transmission mode, the previous literatures and experiences only describe their applicable occasions qualitatively according to the different technical characteristics of AC and DC transmission. But there lacks staff to analyze their applicable range including transmission capacity and transmission distance. For example: According to the technical characteristics of AC and DC transmission mode, only AC transmission mode applies to the transmission project which needs to transfer some power down to the local power grid in the middle of the transmission line. While DC transmission mode applies to interconnecting the two AC networks which operate at non synchronism. However, when the power delivery mode is point-to-net and there is no need to transfer some power down to the local power grid in the middle of the transmission line, like the power delivery of energy bases, AC and DC transmission modes both apply to the occasion. But there is scant literature to study and summarize how to select the transmission mode and its voltage grade on this occasion, based on which this paper will carry out research on applicable range of AC and DC transmission mode and their voltage class sequence.

For the selection of AC transmission voltage grade, the past empirical data and formulas provide some suggestions. But when applying them to the research on the applicable range of AC voltage class sequence, the results are different. Owing to the lack of corresponding power system model, they are unnecessarily suitable for the growing power grid. With the equivalent mathematical model of AC transmission system, the transfer capability of AC transmission line is discussed. It guides the establishment of AC transmission schemes and lays the foundation for the research on the applicable range of AC transmission voltage class sequence.

While for the research on the applicable range of DC transmission voltage class sequence, DC voltage class standardized series and their typical arrangements are determined and applied in China, and the transmission capacity and the economical transmission distances for HVDC transmission systems of each voltage grade are also obtained. It provides reference for the selection of DC transmission scheme in the research on the applicable range of AC and DC transmission mode.

Aiming at the selection of transmission mode and voltage level in the power delivery of energy bases, the paper proposes a comprehensive evaluation method to analyze the applicable range of AC and DC transmission voltage class sequence including transmission capacity and transmission distance, based on the differences in the economic and technical characteristics between AC and DC transmission mode. Firstly, based upon the equivalent mathematical model of AC transmission system, a fast establishing method of AC transmission scheme is designed under the given transmission scenario, and the economical applicable range of EHV and UHV AC transmission is discussed under multiple transmission scenarios. Then, considering many factors such as the power transmission properties, economy and environmental impact, the suitable evaluation index system and grid comprehensive optimum seeking model are built for a research on the applicable range of AC and DC transmission mode. Meanwhile, on the basis of the applicable range of EHV and UHV AC transmission and applicable range of DC transmission voltage class sequence, the suitable AC and DC transmission schemes are selected under the given transmission scenario. And then a grey comprehensive evaluation and analysis can be carried out on them. With multiple transmission scenarios, the applicable range of AC and DC transmission mode and their voltage class sequence are summarized. The final result proves the validity of the proposed research method in this paper.

Under the given transmission scenario, the suitable voltage can be selected referring to the applicable range of AC and DC transmission voltage class sequence. Then the AC and DC transmission scheme are established to meet the power transfer. There are big differences between the two schemes not only in economy but also in power transmission characteristics and reliability. These characteristics play important roles in the selection of AC and DC transmission scheme. Thus we need to choose suitable indexes from these characteristics and then obtain the index values of the two transmission schemes. Evaluate the two schemes comprehensively and the optimal scheme can be selected. Therefore, it has great practical significance for the selection of alternative transmission schemes to establish a set of consummate and reasonable comprehensive evaluation index system for AC and DC transmission mode.

Through the analysis of similarities and differences in the technical economic characteristic between AC and DC transmission, this paper chooses six indexes such as maximum transfer power, annual cost, forced interruption duration, transmission corridor, reactive power loss and fault loss power. These indexes reflect the economy, power transmission characteristics and reliability of AC and DC transmission schemes. They are used to compare the schemes systematically and scientifically. The constructed comprehensive evaluation index system for AC and DC transmission mode is shown in Fig.1.1.

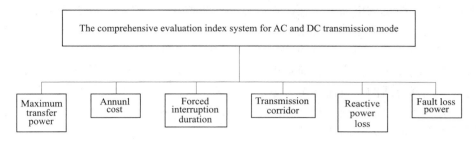

Fig.1.1　The comprehensive evaluation index system for AC and DC transmission mode

第一节　交、直流输电电压等级的适用范围研究

我国能源资源与能源需求呈逆向分布，客观上需要能源在大范围内进行优化配置。现今，我国已经成功建设并投运了特高压交流和直流输电工程，高压输电网中主要形成了 1000/500kV 超特高压交流输电电压序列，以及±500、±660、±800、±1100kV 高压直流输电电压序列。由于交流输电和直流输电在工程投资、环境影响和安全稳定等方面存在差异性，因此，在电网规划建设中，从经济技术特性的角度分析和探讨交直流输电方式的合理适用范围具有重要意义。

对于交直流输电方式的适用范围，现有文献和经验仅仅从两者不同的技术特性来对其各自的适用场合进行定性描述，鲜有文献分析其适用范围，以及输电容量和传输距离。例如，根据交流和直流输电方式的技术特性，只有交流输电方式适用于将线路中间的电能输送到地方电网的输电工程，而直流输电方式则适用于将两个非同步交流电网连接起来的输电工程。然而，在点对点输电方式且在输电线路中间不要求能量传输到地方电网的情况下，交、直流输电方式均适用。但很少有文献研究和总结如何在这种情况下选择输电方式和电压等级。为此，本文将针对该种情况下交、直流输电方式的适用范围和电压等级展开研究。

对于交流输电电压等级的选择，过去的经验数据和公式提供了一些建议，但将它们运用于研究交流电压等级序列时，结果却不一样。由于缺乏相应的电力系统模型，所以它们均不适用于正在发展中的电网。利用交流输电系统的等值数学模型，可以探讨研究交流输电线路的传输容量。该模型能够指导交流输电方案的建立，也可以为交流输电电压等级适用范围的研究奠定基础。

而从直流输电电压等级的适用范围研究来看，目前已经确立了其标准化序列和典型方案，并且在我国已经应用起来，同时也确定了各个电压等级的高压直流输电系统的传输能力和经济传输距离。它可以为交、直流输电方式适用范围研究中的直流输电方案选择问题提供参考。

本文基于交流输电与直流输电两者在经济、技术特性方面的差异，针对能源基地输送需求下输电方式和电压等级的选择问题，提出了一种综合考虑经济、技术因素的方法，用以评估交、直流输电方式所适用的输电容量和输电距离的范围。首先，利用交流输电系统的等值数学模型，设计了给定输电需求下交流输电方案的快速建立方法，并在多种输电需求下讨论了超高压和特高压交流输电的经济适用范围。其次，考虑功率传输特性、经济性、环境影响等多方面因素，建立适宜于交、直流输电适用范围研究的评价指标体系和电网综合优选模型。同时，基于超高压和特高压交流输电的适用范围，以及直流电压等级的适用范围，在给定输电需求下选择合适的交直流输电方案，然后利用灰色综合评价及分析对它们进行研究。在多组输电情景下，总结了交、直流输电方式的适用范围和它们的电压等级序列。最后的仿真结果证明了本文所提方法的有效性。

在给定的输电情况下，可以根据交、直流输电电压等级的适用范围来选择合适的电压，然后为满足电能输送的要求建立交、直流输电方案。两者不仅在经济上，而且在功率传输特性、安全稳定性等特性上存在较大差异。这些特性对交、直流输电方案的决策起着重要作用。所以需要从这些特性中选择合适的指标，获取两种输电方案的指标值，以此对两种输电方案进行综合评价，继而选出最优方案。因此，建立一套完善合理的交、直流输电方式适用范围的综合评价指标体系，对备选输电方案的决策具有重要的现实意义。

通过分析交、直流输电在技术经济特点上的共性和差异，本文从经济性、功率传输特性和安全稳定性等方面选取了最大传输功率、年费用、强迫停运时间、输电走廊、无功损耗和故障损失功率 6 个评价指标，对交、直流输电方案的特性进行较为系统而科学的比较，所构建的交、直流输电综合评价指标体系如图 1.1 所示。

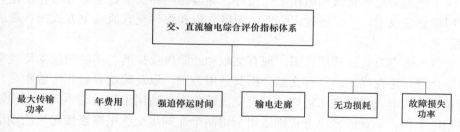

图 1.1 交、直流输电综合评价指标体系

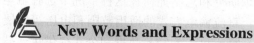

New Words and Expressions

voltage	n. [电] 电压
voltage class sequence	电压等级序列
synchronism	n. 同时发生，同步性
energy base	能源基地
voltage grade	电压等级
forced interruption	强行断流，强行断电
transmission corridor	送电回廊，输电走廊
electrical grid	电网
loss power	损耗功率

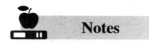
Notes

1. Only AC transmission mode applies to the transmission project which needs to transfer some power down to the local power grid in the middle of the transmission line.

只有交流输电方式适用于将线路中间的电能输送到地方电网的输电工程。

2. It guides the establishment of AC transmission schemes and lays the foundation for the research on the applicable range of AC transmission voltage class sequence.

该模型能够指导交流输电方案的建立，也可以为交流输电电压等级适用范围的研究奠定基础。

3. Aiming at the selection of transmission mode and voltage level in the power delivery of energy bases, the paper proposes a comprehensive evaluation method to analyze the applicable range of AC and DC transmission voltage class sequence including transmission capacity and transmission distance, based on the differences in the economic and technical characteristics between AC and DC transmission mode.

本文基于交流输电与直流输电两者在经济、技术特性方面的差异，针对能源基地输送需求下输电方式和电压等级的选择问题，提出了一种综合考虑经济、技术因素的方法，用以评估交、直流输电方式所适用的输电容量和输电距离的范围。

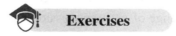
Exercises

Ⅰ. Choose the best answer into the blank.

1. There is scant literature to analyze their applicable range including transmission capacity and transmission ____.
    A. rate   B. mode   C. voltage grade   D. distance

2. When the power delivery mode is ____, AC and DC transmission modes both apply to the occasion.
    A. point-to-net   B. point-to-point   C. net-to-net   D. net-to-point

3. The paper is aiming at the selection of transmission mode and voltage level in the power delivery of ____.
    A. power plant   B. energy bases   C. substation   D. convertor station

4. A fast establishing method of AC transmission scheme is designed under the given transmission ____.
    A. capacity   B. scheme   C. rate   D. scenario

5. The AC and DC transmission ____ are established to meet the power transfer.
    A. mode   B. scheme   C. rate   D. scenario

6. The maximum transfer power reflects the transmission ____ of each established transmission scheme.

A. capacity　　　　B. scenario　　　　C. mode　　　　D. rate

Ⅱ. Answer the following questions according to the text.

1. What does the HV power transmission network mainly consist of?

2. Why are the past empirical data and formulas provide some suggestions unnecessarily suitable for the growing power grid?

3. Are DC voltage class standardized series and their typical arrangements determined and applied in China?

4. What factors should be considered when the suitable evaluation index system and grey comprehensive optimum seeking model is built for research on the applicable range of AC and DC transmission mode?

5. Through the analysis of similarities and differences in the technical economic characteristic between AC and DC transmission, how many indexes does this paper choose, and listed?

Ⅲ. Translate the following into Chinese.

For the selection of AC transmission voltage grade, the past empirical data and formulas provide some suggestions. But when applying them to the research on the applicable range of AC voltage class sequence, the results are different. Owing to the lack of corresponding power system model, they are unnecessarily suitable for the growing power grid. With the equivalent mathematical model of AC transmission system, the transfer capability of AC transmission line is discussed. It guides the establishment of AC transmission schemes and lays the foundation for the research on the applicable range of AC transmission voltage class sequence.

Keys

Ⅰ. Choose the best answer into the blank.

1. C　2. A　3. A　4. B　5. C　6. A

Ⅱ. Answer the following questions according to the text.

1. It consists of the 1000 kV UHV and 500 kV EHV AC transmission voltage class sequence and ±500, ±660, ±800, ±1100 kV HVDC voltage class sequence.

2. Owing to the lack of corresponding power system model, they are unnecessarily suitable for the growing power grid.

3. Yes.

4. Such as the power transmission properties, economy and environmental impact, and so on.

5. Six. They are maximum transfer power, annual cost, forced interruption duration, transmission corridor, reactive power loss and fault loss power.

Ⅲ. Translate the following into Chinese.

对于交流输电电压等级的选择，过去的经验数据和公式提供了一些建议，但将它们运用于研究交流电压等级序列时，结果却不一样。由于缺乏相应的电力系统模型，所以它们均不适用于正在发展中的电网。利用交流输电系统的等值数学模型，可以探讨研究交流输电线路的传输容量。该模型能够指导交流输电方案的建立，也可以为交流输电电压等级适用范围的研究奠定基础。

翻译技巧之专业词汇的构成

在科学技术的各个领域，文献资料中都存在大量的专业词汇（仅用于某一学科或与其他学科通用的词汇或术语）。对专业词汇的快速认读和正确理解，是提高阅读速度和增强理解能力的关键，也是阅读和翻译相关领域专业技术文献的必要和知识积累。

专业词汇的形成主要有三种情况：

（1）借用日常英语词汇或其他学科的专业词汇，但是词义和词性可能发生了明显的变化。例如：在日常英语中表示"力量、权利"和在机械专业表示"动力"的power，在电力专业领域可以仍作为名词，表示"电力、功率、电能"；也可以作为动词，表示"供以电能"。在日常英语中表示"植物"的plant一词，在电力专业领域中用来表示"电厂"等。

（2）由日常英语词汇或其他学科的专业词汇，直接合成新的词汇。例如：over和head组合成overhead，表示"架空（输电线）"；super和conductor组合成superconductor，表示"超导体"等。

over + head = overhead   在……上+头=在头上的，高架的
overhead lines   架空线
super + conductor = superconductor   超+导体=超导体

（3）由基本词根（etyma）和前缀（prefixes）和后缀（suffixes）组成新的词汇。大部分专业词汇属于这种情况。一般来说，加前缀改变词意，加后缀改变词性。

例如：
词根   develop   *vt.* 发达，发展，开发
加后缀-ed   developed   *adj.* 发达的
再加前缀un-   undeveloped   *adj.* 不发达的，未开发的

Section 2 An Overview of Power Transmission Systems in China

1. Overview of the development of China's power industry and power transmission systems

In 1882, the origin of the Shanghai Electrical Company symbolized the beginning of China's power industry. By the end of 1949, the total installed capacity was 1.85GW. After 1949, China's power industry developed much faster. The annual total installed capacity increased by 10%. In the two years, 1987 and 1995, the total capacity exceeded 100GW and 200GW, respectively. In 1996, the total capacity reached to stand at the second place in the world. Since 2000, China's GDP has increased 10% annually; correspondingly, the country's power industry entered into a new prosperous period. The total installed capacity reached 300GW in April 2000, 400GW in May 2004, and 500GW in December 2005. The power generation of 2005 was 2414TWh, at a 13.3% increase of the previous year. In 2006, the installed capacity exceeded 622GW. It is estimated that the development of China's power industry will keep growing in the next 10~20 years. By 2020, the total installed capacity is expected to exceed 1500GW.

China's power grid has passed through several periods. Before 1949, it developed very slowly, with a small scale. By the end of 1949, the total length of 20kV or above transmission lines was 6475km; the total capacity of all transformer substations was 3460MVA. After 1949, the construction of China's power grid accelerated, and the standard for voltage level of transmission lines was established, too. In 1954, the first 220kV power transmission project, the Song Dong Li project in the northeast region of China was completed, thus, transferring power from the Fengman hydro plant to Liaoning province which is the base of China's heavy industry. In 1955, the first 110kV power transmission project, Beijing Dongbeijiao to Guanting project, was completed with transmission lines stretching 105.9km. In 1972, the first 330kV power transmission project, the Liu Tian Guan project, was completed. It started from the Liujiaxia hydro plant to the Tangyu substation in Mei County, Guanzhong, Shanxi province via the Qin'an substation in Tianshui, Gansu province. The length of the transmission lines is 534km. This project established China's Northwest Grid across several provinces. In 1981, the first 500kV power transmission project, Ping Wu project, was completed. The length of the transmission line is 594.8km, which starts from the Yaomeng power plant in Pingdingshan, Henan province to the Fenghuangshan substation in Wuchang, via the Shuanghe substation in Hubei province. Later, 500kV backbone grids were formed gradually in the four regional grids including several provincial grids, i.e., the Central China Gird, the Northeast China Grid, the North China Grid, and the East China Grid. A 330kV backbone grid was formed in the Northwest China Grid. In 1989, the 500kV HVDC project from Gezhouba to Shanghai was completed; the length of the transmission lines is 1052km, and the rated capacity is 1200MW. This project interconnected two regional grids in China for the first time. In September 2005, the first 750kV power transmission project started, which demonstrated the great progress made in China's power transmission technology. In the next 10~20 years, China will continuously fund the construction of ultra high voltage grids, and a nationwide grid with a backbone grid of 1000kV AC lines and 800kV DC lines.

With the drastic increase in the total length of transmission lines and higher voltage levels, China's big regional grids are being interconnected gradually. At the end of 1980s, the completion of the 500kV Gezhouba to Shanghai HVDC project established the trans-regional interconnection. In 2001, the North China Power Grid and the Northeast China Power Grid were interconnected. It was the first time that two big regional grids were interconnected via an AC line in China. In October 2001, a 500kV AC line interconnected the East China Power Grid and Fujian Power Grid. In May 2002, the project to transmit power from Sichuan province to Central China was completed, interconnecting the Chuanyu Power Grid and the Central China Power Grid. On May 5th, 2003, the 500kV DC line with a rated capacity of 3000MW from the Three Gorges to Changzhou was put in to use. This line had the highest transmission capability in the world at that time, transferring power from the Three Gorges hydro plant to East China. In September 2003, the Central China Power Grid and the North China Power Grid were interconnected, and thereby an inter-regional AC synchronized grid including the Northeast China Power Grid, the North China Power Grid, the Central China Power Grid including (the Chuanyu Power Grid) was formed. In 2004, this inter-regional gird was interconnected with the South China Power Grid by a DC line from the Three Gorges to the Guangdong province. In July 2005, a back-to-back DC project interconnected the Northwest China Power Grid and the Central China Power Grid, completing the nationwide grid interconnection. Later, with the completion of the Three Gorge hydro plant and the Three Gorge transmission system, a large AC and DC transmission system including the Central China Power Grid and the East China Power Grid was formed. The forecasted total installed capacity is more than 200GW. In the future, with the completion of the north, middle, and south corridors in the "Power Transmission from West to East" plan, there will be a nationwide interconnected grid including the north, middle, and east interconnected grids in China.

2. HVAC (EHVAC and UHVAC) transmission systems

At present, transmission lines in China's power transmission system can be categorized into four voltage levels, i.e. 220kV, 330kV, 500kV, and 750kV. The 330kV and 500kV transmission lines make up the backbone grids. At the end of 2005, there were 540 500kV lines with a total length of 63 790km and 204 500kV substations with total transformer capacity of 282 100MVA in major power grids. There were 154 330kV lines with total length of 12 842km and 60 330kV substations with total transformer capacity of 25.21GVA in the Northwest China. In the1980s, the Ministry of Electric Power launched a study about higher voltage level. According to the study, 750kV and 1000kV are specified as the standard voltage levels, and they are one level higher than 330kV and 500kV, respectively. On September 26, 2006, the first 750kV AC line in China from Guanting to Lanzhou was put in use. In August 2006, China's first 1000kV AC pilot project, from Changzhi, Shanxi province via Nanyang, Henan province to Jingmen, Hubei province, reached the stage of implementation. The two lines and some other lines that will be constructed in future will constitute China's future backbone grids.

第二节 中国电力传输系统概述

1. 中国电力工业与电力传输系统发展概述

1882 年上海电力公司的成立，标志着中国电力工业的开启。截至 1949 年年底，总装机容量达到 1.85GW。1949 年后，中国的电力工业发展更快，每年的总装机容量增长 10%。在 1987 年和 1995 年这两年中，总装机容量分别超过了 100GW 和 200GW。1996 年，总装机容量已高居世界第二位。自 2000 年以来，中国的国内生产总值年均增长 10%，中国的电力工业也因此而进入了一个新的繁荣时期,总装机容量在2000年4月、2004年5月、2005年12月分别达到300、400、500GW。2005年的发电量为2414TWh,相比于前一年增加了13.3%。在2006年,总装机容量超过了622GW。据估计,在未来10~20年,中国电力工业将持续增长。到2020年,装机总容量将有望超过1500GW。

中国的电网经历了几个发展阶段。在1949年以前,电网的发展速度很慢,且规模较小。截至1949年年底,20kV及以上输电线路的总长度为6475km,所有变电站的总容量为3460MVA。1949年以后,中国的电网建设加快了步伐,并建立了输电线路电压等级标准。1954年,首个220kV输电工程——位于中国东北地区的松东李工程竣工,从而将从丰满水电厂生产的电能输送到中国的重工业基地辽宁省。1955年,首个110kV输电工程——北京东北角至官厅工程竣工,其输电线路长达105.9km。1972年,首个330kV输电工程——刘天关工程——竣工,这条线路始于刘家峡水电厂,途经甘肃天水秦安变电站,最后止于陕西关中眉县汤峪变电站,其建立了横跨多个省的西北电网,总长达534km。1981年,首个500kV输电工程——平武工程——竣工,该线路始于河南平顶山姚孟电厂,经湖北双河变电站,最后止于武昌凤凰山变电站,其输电线路长达594.8km。随后,500kV骨干网逐渐在华中电网、东北电网、华北电网、华东电网等四个由多个省级电网组成的区域电网中被建立起来。其中,在西北电网中建立了一个330kV的骨干网。1989年,500kV葛洲坝到上海高压直流输电工程竣工,其输电线路长达1052km,额定容量为1200MW,首次实现了中国两个区域电网的互联。2005年9月,首个750kV输电工程的启动标志着中国在输电技术领域取得了长足进步。在接下来的10~20年内,中国将持续投资建设特高压电网,以及一个包含1000kV交流线路、800kV直流线路骨干网的全国电网。

随着输电线路总长度和高电压等级的急剧增加,中国的大区域电网正在逐步实现互联。在20世纪80年代末,500kV葛洲坝到上海高压直流输电工程的完工实现了跨区域互联。2001年,华北电网与东北电网实现了互联,这是中国首次通过交流输电线路实现两个大区域电网互联。2001年10月,一条500kV交流输电线路将华东电网和福建电网连接起来,而2002年5月建成的从四川至华中的输电工程实现了川渝电网和华中电网的互联。2003年5月5日,从三峡到常州的额定容量为3000MW的500kV直流线路正式投入使用,这条将电能从三峡水电站输送到华东地区的线路具有当时世界上最高的输电容量。2003年9月,华中电网和华北电网实现互联,从而形成了包含东北电网、华北电网、华中电网(包括川渝电网)的跨区域交流同步电网。2004年,从三峡到广东的直流输电线路实现了该跨区域电网与中国南方电网的互联。2005年7月,一项背靠背直流输电工程将西北电网与华中电网连接起来,这也标志着全国性互联电网的完成。此后,随着三峡水电站和三峡输电系统的建成,一个包含华中电网、华东电网在内的交、直流输电系统也随之形成,其预计总装机容量超过200GW。未来随着"西电东送"计划中的北方、中部和南方输电走廊的建成,将形成包括北方、中部及东部互联电网的全国性互联电网。

2. 高压交流(超高压交流和特高压交流)输电系统

目前,中国输电系统中的输电线路按电压等级划分可分为四种:220、330、500、750kV,其中330、500kV输电线路是骨干网。截至2005年年底,主网中共有540条总长度为63 790km的500kV线路,以及204个总容量为282 100MVA的500kV变电站;在中国的西北部,有154条总长达12 842km的330kV线路,以及60个总容量为25.21GVA的330kV变电站。在20世纪80年代,电力部开始着手了一项关于更高电压等级的研究。根据这项研究,750kV

和 1000kV 被指定为标准电压等级，它们分别比 330kV 和 500kV 高一个等级。2006 年 9 月 26 日，中国首条从官厅到兰州的 750kV 交流输电线路正式投入使用。2006 年 8 月，中国首个从山西长治，经河南南阳再到湖北荆门的 1000kV 交流试点项目步入了实施阶段。这两条线路和未来建成的其他线路将构成中国电网未来的骨干网架。

New Words and Expressions

transformer	*n.* [电] 变压器
substation	*n.* 变电站
backbone	*n.* 支柱；主干网
interconnected	*adj.* 连通的；有联系的
total installed capacity	总装机容量
power generation	发电，发电量；发电总设备
power grid	电力网
voltage level	电压等级；电压水平
transfer power	转化功率，转换功率
power plant	发电厂
backbone grid	骨干网架，骨干网络
regional grid	地区电网；区域电网
provincial grid	省级电网；省电网
rate capacity	额定载荷；额定容量；额定量

Notes

1. In 1996, the total capacity reached to stand at the second place in the world.
1996 年，总装机容量居世界第二位。

2. It is estimated that the development of China's power industry will keep growing in the next 10～20 years.
据估计，在未来 10～20 年，中国电力工业将持续增长。

3. In September 2005, the first 750kV power transmission project started, which demonstrated the great progress made in China's power transmission technology.
2005 年 9 月，首个 750kV 输电工程的启动标志着中国在输电技术领域取得了长足进步。

Exercises

Ⅰ. Choose the best answer into the blank.

1. The total installed capacity reached 500GW in _____?
   A. April 2000　　　B. May 2004　　　C. December 2005　　D. Novmber 2007

2. _____ 1949, the standard for voltage level of transmission lines was established.

A. Before B. In C. By the end of D. After

3. The 500kV HVDC project interconnected _____ regional grids in China for the first time.

A. two B. three C. four D. five

4. The line from the Three Gorges to Changzhou had the _____ transmission capability in the world at that time.

A. biggest B. greatest C. largest D. highest

5. In the future, with the completion of the _____ corridors in the "Power Transmission from West to East" plan, there will be a nationwide interconnected grid.

A. west, middle, and east
B. west, middle, and south
C. north, middle, and south
D. north, middle, and east

6. The _____ and _____ transmission lines make up the backbone grids.

A. 220kV; 330kV B. 330 kV; 500kV C. 330kV; 750kV D. 500kV; 750kV

II. Answer the following questions according to the text.

1. What did symbolize the beginning of China's power industry?
2. Which is the first 330kV power transmission project in China?
3. How many 500kV backbone grids were formed gradually in the four regional grids, and listed?
4. What urged China's big regional grids to be interconnected gradually?
5. How many transmission voltage levels in China?

III. Translate the following into Chinese.

In September 2003, the Central China Power Grid and the North China Power Grid were interconnected, and thereby an inter-regional AC synchronized grid including the Northeast China Power Grid, the North China Power Grid, the Central China Power Grid including (the Chuanyu Power Grid) was formed. In 2004, this inter-regional gird was interconnected with the South China Power Grid by a DC line from the Three Gorges to the Guangdong province. In July 2005, a back-to-back DC project interconnected the Northwest China Power Grid and the Central China Power Grid, completing the nationwide grid interconnection. Later, with the completion of the Three Gorge hydro plant and the Three Gorge transmission system, a large AC and DC transmission system including the Central China Power Grid and the East China Power Grid was formed. The forecasted total installed capacity is more than 200GW. In the future, with the completion of the north, middle, and south corridors in the "Power Transmission from West to East" plan, there will be a nationwide interconnected grid including the north, middle, and east interconnected grids in China.

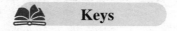

Keys

I. Choose the best answer into the blank.

1. C   2. A   3. A   4. B   5. C   6. A

II. Answer the following questions according to the text.

1. The origin of the Shanghai Electrical Company.

2. The Liu Tian Guan project.

3. Four.They are the Central China Gird, the Northeast China Grid, the North China Grid, and the East China Grid.

4. With the drastic increase in the total length of transmission lines and higher voltage levels.

5. Four.

Ⅲ. Translate the following into Chinese.

2003年9月，华中电网和华北电网实现了互联，从而形成了包含东北电网、华北电网、华中电网（包括川渝电网）的跨区域交流同步电网。2004年，从三峡到广东的直流输电线路实现了该跨区域电网与中国南方电网的互联。2005年7月，一项背靠背直流输电工程将西北电网与华中电网连接起来，这也标志着全国性互联电网的完成。此后，随着三峡水电站和三峡输电系统的建成，一个包含华中电网、华东电网在内的交、直流输电系统也随之形成，其预计总装机容量超过200GW。未来随着"西电东送"计划中的北方、中部和南方输电走廊的建成，将形成包括北方、中部及东部互联电网的全国性互联电网。

翻译技巧之词类转换

由于英汉两种语言在语法和习惯上存在差异，在描述同一事物时，两种语言会使用不同的词性。因此，在翻译过程中，有时需要变换原词的词性，以有效地传达原文的信息。常见的词类转换有以下几种。

1. 英语名词→汉语动词

英语中具有动作意义或由动词转化过来的名词，汉译时往往转化为动词。如：

His arrival at this conclusion was the result of much thought. 他得出这结论是深思熟虑的结果。

2. 英语动词→汉语名词

有些动词在汉语里虽有相对应的动词词义，但在某种特定的语言场合下不能使用该动词词义，或由于需要选择另一种更好的译文，因而将其转化为汉语的名词。如：

Judgment should be based on facts, not on hearsay. 判断应该以事实为依据，而不应该凭借道听途说。

The man I saw at the party looked and talked like an American. 我在晚会上看到的那个人，外表和谈吐像美国人。

The electronic computer is chiefly characterized by its accurate and rapid computation. 计算机的主要特点是计算准确迅速。

3. 英语名词→汉语形容词

英语原文中有形容词加后缀构成的名词，翻译时可转化为汉语的形容词。如：

Scientists are no strangers to politics. 科学家对政治并不陌生。

We found difficulty in solving the complicated problem. 我们发现解决这个复杂的问题是

很困难的。

4. 英语介词副词→汉语动词

He came to my home for help. 他来到我家请求帮助。
I love having Friday off. 我喜欢周五休息。
I am for the former. 我支持前者。
I am against the latter. 我反对后者。

5. 英语副词→汉语名词

It was officially announced that Paris is invited to the meeting. 官方宣布，巴黎应邀出席会议。

Section 3 HVDC Transmission Systems

HVDC power transmission technology is an effective supplement of AC power transmission technology. HVDC technology can be applied in long distance and bulk power transmission, allowing the interconnection of large regional AC power systems and power transmission to islands through seabed cables. In China's power grid plan, the HVDC technology will play a very important role. It will be used to interconnect several large power grids for transmitting hydroelectric and thermal power from northwest and southwest China to more developed coastal regions in the southeast, and transmit bulk power generated by the Three Gorges Hydro Plant to China's other areas. At present, there are six 500kV HVDC projects in China, i.e., the 1200MW Gezhouba-Nanqiao project, the 3000MW Three Gorges-Changzhou project, the 3000MW Three Gorges-Guangdong project, the 1800MW Tianshengqiao-Guangdong project, the 3000MW Guizhou-Guangdong project, and the 3000MW Three Gorges-Shanghai project. Additionally, the second 3000MW Guizhou-Guangdong DC line will be put in operation in 2007, and the 800kV Yunnan-Guangdong and other four ultra HVDC projects are under construction or in the planning stage.

1.1 The 500kV HVDC transmission in the Three Gorges power system

The Three Gorges power transmission system is built for transferring the bulk power from the Three Gorges Hydro Plant to east China, central China and south China. Several HVDC lines are included in the system. By now, 2965km overhead DC lines and six converter substations with a total capacity of 18 000MW have been built. Several important HVDC projects will be introduced as the following.

1. Three Gorges-Changzhou HVDC transmission

The project is the first 500kV, 3000MW HVDC project in China and also the second interconnection project between the Central China Power Grid and the East China Power Grid (the first is the Gezhouba-Shanghai HVDC project). The project aims to transfer power from the Three Gorges Hydro Plant to East China power grids. The length of the transmission line is 890km, from the Longquan converter station to the Zhengping converter station. In 2003, the project was put into operation.

2. Three Gorges-Guangzhou HVDC transmission

The project delivers power from the Three Gorges Hydropower Plant in Central China to the load center located in the Guangdong province, enhancing asynchronous link for the two power systems. The 940km long transmission line went into commercial operation in 2004. One of the HVDC converter stations is located at Jingzhou near the Three Gorges Power Plant, and the other in Huizhou near Guangzhou.

3. Three Gorges-Shanghai HVDC transmission

The project is the fourth largest HVDC transmission with rated power of 3000MW, and also the largest power transmission project during the period of the 11th Five Year Plan of the State Grid Corporation of China. One HVDC converter station is located at Yidu, approximately 58km from the Three Gorges power plant, and the other in Huaxin at the west outskirt of Shanghai. The receiving station will feed power directly into the Shanghai 500kV AC ring. In 2006, the 1060km power link went into commercial operation.

1.2 The 500kV HVDC transmission in the Southern China Power Grid

The HVDC transmission systems in the South China Power Grid are an important part of the "Power Transmission from West to East" project, within which several HVDC projects in operation are as the following.

1. Tianshengqiao-Guangzhou HVDC transmission

The project is a part of the "Power Transmission from West to East" project. It aims to transfer abundant hydropower from the Tianshengqiao Hydroelectric Power Plant, Yunnan and Guizhou to Guangzhou which is a big load center in South China. The DC voltage, rated current, and rated power of the project are 500kV, 1800A, and 1800MW, respectively. The project includes two converter substations. One is located in Mawo near the Tianshengqiao power plant, and the other is located in the northern outskirts of Guangzhou. The length of the transmission line between the two converter substations is 960km. The project went into operation in 2001.

2. Guizhou-Guangdong HVDC transmission

The project aims to transmit abundant hydroelectric and thermal power from southwest China to the load center in the delta area of the Zhujiang River. The DC voltage, rated current, and rated power of the project are 500kV, 3000A, and 3000MW, respectively. The project includes two converter substations. One of them is located in Anshun in the Guizhou province, and the other is located in Zhaoqing in the Guangdong province. The length of the transmission line between the two converter substations is 880km. The project went into operation in 2004.

3. The second Guizhou-Guangdong HVDC transmission

This HVDC transmission is a backbone project of the "Power Transmission from West to East project". The DC voltage, rated current, and rated power of the project are 500kV, 3000A, and 3000MW, respectively. The project includes two converter substations. One of them is located in Xingren in the Guizhou province, and the other is located in Shenzhen in the

Guangdong province. The length of the transmission line between the two converter substations is 1225km. The project will go into operation in 2007.

1.3 The 800kV ultra high voltage DC transmission projects in South China grid and the Jinsha River transmission system

The 800kV or above DC transmission can be categorized as ultra HVDC transmission, which can be used to transmit bulk power over long distance from big hydroelectric power plants or thermal power plants to load centers, as well as interconnect two asynchronous power grids. Now, 600kV is the highest voltage of all HVDC transmission lines in operation in the world. At present, several ultra HVDC lines are under construction in China:

1. Yunnan-Guangdong 800kV ultra HVDC transmission

This project will be the first 800kV 5000MW ultra HVDC transmission in the world. The 1446km overhead DC line begins at the Lufeng converter station in the Yunnan province and ends at the Zengcheng converter station in the Guangdong province. This DC line will have the largest transmission capacity in China, and it will go into commercial operation in July 2010.

2. Jinsha River 800kV ultra HVDC transmission

The upstream of the Changjiang River from the Yushu in the Qinghai province to Yibing in the Sichuan province is called the Jinsha River. The river has a length of 2360km and a fall of 3280m which brings abundant water energy resource. Several hydropower plants will be built on the river. The Xiluodu and Xiangjiaba hydro plants which have rated capacity of 18 600GW, larger than the Three Gorges Hydro Plant. The Xiangjiaba-Shanghai ultra HVDC transmission line is a part of the Jinsha River transmission system. The DC voltage and the rated capacity of the project are 800kV and 6400MW, respectively. The 2034km overhead DC line begins at the Fulong converter station in the Sichuang province and ends at the Fengxian converter station in Shanghai. It is expected to go into commercial operation in July 2012.

1.4 Future development

Most fossil and water energy resources are located in the west of China; however, major load centers are located in the middle and the east of China, making power transfer from the west to the east very important to China's economic growth. Three transmission corridors, i.e., the north, the middle, and the south corridors, will be built. In these corridors, more than 20 HVDC lines will be built in the next 10~15 years.

第三节 高压直流输电系统

高压直流输电技术作为交流输电技术的有效补充，适用于长距离、大功率输电，可以实现大区域交流电力系统之间的互联，以及通过海底电缆向岛屿输电的场合。在中国的电网建设规划中，高压直流输电技术将会扮演非常重要的角色。该技术将会被运用于几个大型电网之间的互联，从而将西北和西南的水电、火电输送给东南较为发达的沿海地区，同时也可以将三峡水电站的电能输送到中国其他地区。目前，中国有 6 个 500kV 高压直流工程，分别是容量为 1200MW 的葛洲坝—南桥工程、3000MW 的三峡—广州工程、1800MW 的天生桥—广东工程、3000MW 的贵州—广东工程以及 3000MW 的三峡—广东工程。除此之外，第二个容量为 3000MW 的贵州—广东高压直流输电线路将会在 2007 年投入使用，而 800kV 的云南—广东及其他四个特高压直流工程正在建设或规划当中。

1.1 三峡电力系统中的 500kV 高压直流输电工程

三峡输电系统是为将三峡水电站的庞大电能输送到中国东部、中部和南部而建立起来的，其包含多条高压直流输电线路。迄今为止，在三峡电力系统中已经建造了总长为 2965km 的架空直流输电线路和 6 个总容量为 18 000MW 的换流站。一些重要的高压直流项目的介绍如下：

1. 三峡—常州高压直流输电工程

该工程是中国首个 500kV、3000MW 输电工程，同时也是第二个实现华中电网与华东电网互联的工程（第一个是葛洲坝—上海高压直流工程），其主要是为了将三峡水电站的电能输送到华东电网。它的输电线路总长为 890km，始于龙泉换流站，止于郑平换流站。该工程已在 2003 年投入使用。

2. 三峡—广州高压直流输电工程

该工程将华中三峡水电站的电能输送到广东省的负荷中心，并增强了这两个电力系统的异步联系。这条 940km 长的输电线路于 2004 年投入了商业使用。该工程当中的一个高压直流换流站坐落于三峡电厂附近的荆州，另一个在广州的惠州。

3. 三峡—上海高压直流输电工程

该工程是第四大额定功率为 3000MW 的高压直流输电工程，同时也是国家电网公司"十一五计划"期间最大的输电工程。其中一个高压直流换流站位于距三峡电厂 58km 的宜都，另一个在上海西郊的华新。接收站将会直接向上海 500kV 交流环网提供电能。这条 1060km 输电线路于 2006 年投入了商业运营。

1.2 南方电网 500kV 高压直流输电工程

南方电网的高压直流输电系统是"西电东送"工程的重要一环。其中一些正在运行当中的高压直流工程介绍如下：

1. 天生桥—广州高压直流输电工程

该工程是"西电东送"工程的一部分，主要是将天生桥水电站、云南和贵州的电能输送到中国南部最大的负荷中心——广州。该工程的直流电压、额定电流和额定功率分别为 500kV、1800A、1800MW，其包含两个换流站：一个位于天生桥附近的马窝，另一个位于广州的北郊。这两个换流站之间的输电线路长度为 960km。该工程于 2001 年投入运行。

2. 贵州—广东高压直流输电工程

该工程旨在将中国西南地区丰富的水电和火电输送到珠三角地区的负荷中心，其直流电压、额定电流和额定功率分别为 500kV、3000A、3000MW。该工程包含两个换流站：其中一个在贵州安顺，另一个在广东肇庆。这两个换流站之间的输电线路长度为 880km。该工程于 2004 年投入运行。

3. 第二条贵州—广东高压直流输电工程

该高压直流输电工程是"西电东输"工程的主干工程，其直流电压、额定电流和额定功率分别为 500kV、3000A、3000MW。同样，该工程包含两个换流站：一个坐落于贵州兴仁，另一个坐落于广东深圳。这两个换流站之间的输电线路总长 1225km。该工程将会在 2007 年投入运行。

1.3 南方电网和金沙江输电系统中的 800kV 特高压直流输电工程

800kV 及以上直流输电可以归为特高压直流输电范畴，用于将大型水电站或火电厂与负荷中心之间的大功率、长距离电能传输，同时可以实现两个异步电网之间的互联。现如今，全世界运行当中的高压直流输电线路的最高电压等级为 600kV，而目前中国有多个特高压直流输电线路正在建设当中。

1. 云南—广东 800kV 特高压直流输电工程

该工程将会成为世界首个 800kV、5000MW 的特高压直流输电工程，其 1446km 长的架空直流线路从云南禄丰换流站出发，止于广东增城换流站。这条直流线路的传输容量为国内之最，并将于 2010 年 7 月投入商业运营。

2. 金沙江 800kV 特高压直流输电工程

从青海玉树到四川宜宾的长江上游段称为金沙江，其总长为 2360km，落差达 3280m，因此蕴含着丰富的水资源，多个水电站将会在该流域上被建立起来。拥有 18 600GW 额定容量的溪洛渡和向家坝水电站比三峡水电站还大，且向家坝—上海特高压直流输电线路是金沙江输电系统的一部分。该工程的直流电压和额定容量分别为 800kV、6400MW，其直流输电线路始于四川复龙换流站，止于上海奉贤换流站，总长为 2034km，将在 2012 年 7 月投入商业运营。

1.4 未来发展

中国的化石能源和水资源主要集中在西部，而大负荷主要集中在中部和东部，因此，"西电东送"对中国的经济发展至关重要。北部、中部和南部三个输电走廊将会逐步建成，在未来 10~15 年内，这些输电走廊中将会建成超过 20 条的高压直流输电线路。

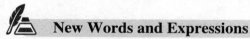

New Words and Expressions

seabed cable	海底电缆
hydroelectric power	水力发电；水电力
thermal power	火力；火力发电；[热] 热功率；热电站
Three Gorges Hydro Plant	三峡水电厂
asynchronous	*adj.* [电] 异步的；不同时的；不同期的
operation	*n.* 运行

Notes

1. It will be used to interconnect several large power grids for transmitting hydroelectric and thermal power from northwest and southwest China to more developed coastal regions in the southeast, and transmit bulk power generated by the Three Gorges Hydro Plant to China's other areas.

该技术将会被运用于几个大型电网之间的互联，从而将西北和西南的水电、火电输送给东南较为发达的沿海地区，同时也可以将三峡水电站的电能输送到中国其他地区。

2. By now, 2965km overhead DC lines and six converter substations with a total capacity of 18 000MW have been built.

迄今为止，在三峡电力系统中已经建造了总长为 2965km 的架空直流输电线路和 6 个总容量为 18 000MW 的换流站。

Exercises

Ⅰ. Choose the best answer into the blank.

1. HVDC technology can be applied in long distance and ____ power transmission.
   A. bulk　　　　　　B. great　　　　　　C. huge　　　　　　D. big

2. ____ project is the first 500kV, 3000MW HVDC project in China.
   A. Three Gorges-Changzhou HVDC transmission
   B. Three Gorges-Guangzhou HVDC transmission
   C. Three Gorges-Shanghai HVDC transmission
   D. Tianshengqiao-Guangzhou HVDC transmission

3. ____ is the largest power transmission project during the period of the 11th Five Year Plan of the State Grid Corporation of China.
   A. Three Gorges-Changzhou HVDC transmission
   B. Three Gorges-Guangzhou HVDC transmission
   C. Three Gorges-Shanghai HVDC transmission
   D. Tianshengqiao-Guangzhou HVDC transmission

4. Which project is not a part of the "Power Transmission from West to East" project?
   A. Three Gorges-Guangzhou HVDC transmission
   B. Tianshengqiao-Guangzhou HVDC transmission
   C. Guizhou-Guangdong HVDC transmission
   D. The second Guizhou-Guangdong HVDC transmission

5. The ____kV or above DC transmission can be categorized as ultra HVDC transmission?
   A. 500　　　　　　B. 800　　　　　　C. 1000　　　　　　D. 1200

6. ____ has a DC line begins at the Lufeng converter station in the Yunnan province and ends at the Zengcheng converter station in the Guangdong province.
   A. Three Gorges-Guangzhou HVDC transmission
   B. Three Gorges-Shanghai HVDC transmission
   C. The second Guizhou-Guangdong HVDC transmission
   D. Yunnan-Guangdong 800 kV ultra HVDC transmission

Ⅱ. Answer the following question according to the text.

1. How many HVDC project are there in China?
2. What is the first interconnection project between the Central China Power Grid and the

East China Power Grid?

3. Which project is the backbone project of the "Power Transmission from West to East" project?

4. Which DC line will have the largest transmission capacity in China?

5. Why will many hydropower plants be built on the Jinsha river?

Ⅲ. Translate the following into Chinese.

HVDC power transmission technology is an effective supplement of AC power transmission technology. HVDC technology can be applied in long distance and bulk power transmission, allowing the interconnection of large regional AC power systems and power transmission to islands through seabed cables. In China's power grid plan, the HVDC technology will play a very important role. It will be used to interconnect several large power grids for transmitting hydroelectric and thermal power from northwest and southwest China to more developed coastal regions in the southeast, and transmit bulk power generated by the Three Gorges Hydro Plant to China's other areas.

Keys

Ⅰ. Choose the best answer into the blank.
1. A   2. A   3. C   4. A   5. B   6. D

Ⅱ. Answer the following questions according to the text.
1. Six.
2. Gezhouba-Shanghai HVDC project.
3. The second Guizhou-Guangdong HVDC transmission.
4. Yunnan-Guangdong 800 kV ultra HVDC transmission.
5. The river has a length of 2360km and a fall of 3280m which brings abundant water energy resource.

Ⅲ. Translate the following into Chinese.

高压直流输电技术作为交流输电技术的有效补充，适用于长距离、大功率输电，可以实现大区域交流电力系统之间的互联，以及通过海底电缆向岛屿输电的场合。在中国的电网建设规划中，高压直流输电技术将会扮演非常重要的角色。该技术将会被运用于几个大型电网之间的互联，从而将西北和西南的水电、火电输送给东南较为发达的沿海地区，同时也可以将三峡水电站的电能输送到中国其他地区。

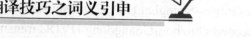

词义引申是指改变原文字面意义的翻译方法。词义引申时，往往可以从词义专业化、词义具体化、词义抽象化和词的搭配四个方面来考虑。

1. 词义专业化

翻译时，遇到一些无法直译或不宜直译的词或词组，应根据上下文和逻辑关系，引申转译。

基于基本词义，根据所涉及专业引申出其专业化语义，以符合技术语言规范和习惯。例如：

Like any precision machine, the lathe requires careful treatment. 与其他精密机械一样，车床需要谨慎使用。

Solar energy seems to offer more hope than any other source of energy. 太阳能似乎比其他任何能源都更有发展前景。

These experiments have produced some valuable. 这些试验得出了某些有价值的数据。

The speed of a plane is associated with the shape of its body and wings. 飞机的速度与机身及机翼的形状有关。

2. 词义具体化

翻译时，根据汉语的表达习惯，把原文中某些词义较笼统的词引申为词义较具体的词，避免译文概念不清或不符汉语表达习惯。例如：

The purpose of a driller is to cut holes. 钻床的功能是加工孔。

At present coal is the most common food of a steam power plant. 目前，煤仍然是凝汽式电厂最常用的能源。

The specific demands that a government wants to make of its scientific establishment cannot generally be foreseen in detail. 一般来说，一国政府对其科研机构提出的具体要求是无法详细预测的。

This nuclear power plant which is computer-controlled will serve the entire city. 这座由电子计算机控制的核电站将向全市供电。

3. 词义抽象化

把原文中词义较为具体的词引申为词义较抽象词，或把词义较形象的词引申为词义较一般的词。例如：

High-speed grinding does not know this disadvantage. 高速磨削不具备这个缺点。

VLSI is still in its infancy. 超大规模集成电路仍处于发展初期。

Elegant systems are in principle unable to deal with…… 完美的理论体系……

Optics technology is one of the most sensational developments in recent years. 光学技术是近年来轰动一时的科学成就之一。

The design of the machine has been improved to answer the challenge of heavy duty climb milling. 机械设计已经得到改进用以应对强力顺铣带来的问题。

4. 词的搭配

遇到动词、形容词与名词搭配时，应根据汉语的搭配习惯，不应受原文字面意义束缚。例如：

hot material 强放射性物质

a leader of the new school 新学派的一位领袖人物

Heat moves from a hotter to a colder body. 热量从温度较高的物体传到温度较低的物体。

Thick liquids pour much slowly than thin liquids. 黏稠的液体比稀薄的液体流动慢得多。

Chapter 2 Mechanics of Overhead Power Transmission Line

Section 1 Introduction to Mechanics of Materials

Mechanics of materials is a branch of applied mechanics that deals with the behavior of solid bodies subjected to various types of loading. It is a field of study that is known by a variety of names, including "strength of materials" and "mechanics of deformable bodies". The solid bodies include axially-loaded bars, shafts, beams, and columns, as well as structures that are assemblies of these components. Usually the objective of our analysis will be the determination of the stresses, strains, and deformations produced by the loads; if these quantities can be found for all values of load up to the failure load, then we will have obtained a complete picture of the mechanical behavior of the body.

Theoretical analyses and experimental results have equally important roles in the study of mechanics of materials. On many occasions we will make logical derivations to obtain formulas and equations for predicting mechanical behavior, but at the same time we must recognize that these formulas cannot be used in a realistic way unless certain properties of the material are known. These properties are available to us only after suitable experiments have been made in the laboratory. Also, many problems of importance in engineering cannot be handled efficiently by theoretical means, and experimental measurements become a practical necessity. The historical development of mechanics of materials is a fascinating blend of both theory and experiment, with experiments pointing the way to useful results in some instances and with theory doing so in others. Such famous men as Leonardo da Vinci (1452—1519) and Galileo Galilei (1564—1642) made experiments to determine the strength of wires, bars, and beams, although they did not develop any adequate theories (by today's standards) to their test results. By contrast, the famous mathematician Leonhard Euler (1707—1783) developed the mathematical theory of columns and calculated the critical load of a column in 1744, long before any experimental evidence existed to show the significance of his results. Thus, Euler's theoretical results remained unused for many years, although today they form the basis of column theory.

The concepts of stress and strain can be illustrated in an elementary way by considering the extension of prismatic bar. A prismatic bar is one that has constant cross section throughout its length and a straight axis. The bar is assumed to be loaded at its ends by axial forces P that produce a uniform stretching, or tension, of the bar. By making an artificial cut (section mm) though the bar at right angles to its axis, we can isolate part of the bar as a free body. At the right-hand end the tensile force P is applied, and at the other end there are forces representing the removed portion of the bar upon the part that remains. These forces will be continuously distributed over the cross section, analogous to the continuous distribution of hydrostatic pressure over a submerged surface. The intensity of force, that is, the per unit area, is called the stress and is commonly denoted by the Greek letter Assuming that the stress has a uniform distribution over the cross section, we can readily see that its resultant is equal to the intensity σ times the cross-sectional area A of the bar. From the equilibrium of the body, we can get this resultant that must be equal in magnitude and opposite in direction to the force P.

$$\sigma = \frac{P}{A} \tag{2.1}$$

as the equation for the uniform stress in a prismatic bar. This equation shows that stress has units of force divided by area—for example, Newton's per square millimeter (N/mm^2) or pounds per square inch (psi).When the bar is being stretched by the forces P, as shown in the figure, the resulting stress is a tensile stress; if the forces are reversed in direction, causing the bar to be compressed, they are called compressive stresses.

A necessary condition for Eq. (2.1) to be valid is that the stress σ must be uniform over the cross section of the bar. This condition will be realized if the axial force P acts through the centroid of the cross section, as can be demonstrated by statics. When the load P does not act at the centroid, bending of the bar will result, and a more complicated analysis is necessary. However, it is assumed that all axial forces are applied at the centroid of the cross section unless specifically stated to the contrary. Also, unless stated otherwise, it is generally assumed that the weight of the object itself is neglected.

原文翻译

第一节 材料力学导论

材料力学是应用力学的一个分支，用来研究固体在各种类型荷载作用下所产生的反应。这个研究领域因许多称谓，如"材料强度""变形固体力学"而被人们所熟知。其研究的固体包括受轴向载荷的杆、轴、梁、圆柱及由这些构件组成的结构。通常，研究的目的在于确定由荷载引起的应力、应变和变形等物理量；如果当固体所承受的荷载达到破坏载荷时，可求得这些物理量，则能够完整地画出该固体的机械性能图。

在材料力学的研究中，理论分析和实验结果同等重要。在许多情况下，我们将进行理论推导，从而得到能够预测材料力学特性的公式和方程。但与此同时，我们必须意识到，除非材料的某些特征是已知的，否则这些公式不能运用于现实情况当中，且这些特征只有在实验室中进行相关的验证后才能为我们所用。同样的，很多重要的工程问题不能用理论的方式来进行有效处理，而通过实验的方法却是一种有效手段。材料力学的历史发展过程是一个具有魅力的理论与实验的结合体，在某些情况下是实验指引了通往获取实用结果的道路，而在另一些情况下却是理论。例如，著名的达芬奇（1452—1519年）和伽利略（1564—1642年）通过做实验测定了钢丝、杆、梁的强度，虽然当时他们的测试结果没有任何准确的理论支撑

（以现在的标准）。相反的，著名数学家欧拉（1707—1783 年）在 1744 年提出了柱体的数学理论，并计算了一个柱体的极限载荷；该理论结果的正确性在很久之后才通过实验证明。虽然现在欧拉的理论结果是圆柱理论的基础，但它们却在很多年里仍未被采用。

  通过考虑等截面杆的拉伸，应力及应变的概念可以用一个基本的方式演示。等截面杆是一个沿长度和直线轴方向具有恒定截面的杆，假设在该杆的末端施加一个产生均匀伸展或拉伸的轴向力 P。沿垂直于轴线的方向将该杆切割后，则可以把杆的各分离部分视为自由体。当张力 P 作用于杆的右端，而代表各分离部分的力作用于其他端点时，这些力将会沿横截面连续分布，类似于作用在被淹没物体表面的连续流体静压分布。力的强度，即单位面积上力的大小，称为应力，并通常用希腊字母表示，且假设压力沿横截面均匀分布。易知，压力的大小等于强度 σ 乘以杆的横截面积 A；由杆力的平衡可知，它与张力 P 的大小相等，方向相反。

$$\sigma = \frac{P}{A} \tag{2.1}$$

上式为等截面杆中均匀应力的计算公式。从这个公式可以看出，应力的单位是力除以面积的单位，即牛顿每平方毫米（N/mm^2）或磅每平方英寸（psi）。当杆在力 P 的作用下被拉伸时，所产生的应力称为拉应力；当施加相反方向的力时杆被压缩，这时所产生的应力称为压应力。

  式（2.1）成立的一个必要条件是应力 σ 必须沿杆的横截面均匀分布。如果轴向力 P 作用于横截面的质心，则该条件自动满足，并通过静力学的验证；当载荷 P 不作用于质心，杆将会产生挠度，因此就需要更加复杂的分析了。但是，如果没有特殊说明，都假定所有的轴向力作用在横截面的质心；同样，除非另有说明，否则物体本身的质量一般是忽略不计的。

New Words and Expressions

mechanics	n.	力学
deformable	a.	可变的
assembly	n.	装配
deformation	n.	变形
strain	n.	张力
blend	n.	混合
equilibrium	n.	平衡
compressive	a.	压缩的
axial	a.	轴向的
centroid	n.	图心
loaded bar		重载梁
prismatic bar		等截面杆
wire	n.	电线
	vt.	给……装电线

Chapter 2 Mechanics of Overhead Power Transmission Line

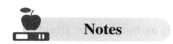
Notes

1. If these quantities can be found for all values of load up to the failure load, then we will have obtained a complete picture of the mechanical behavior of the body.

如果当固体所承受的荷载达到破坏载荷时，可求得这些物理量，则能够完整地画出该固体的机械性能图。

2. The historical development of mechanics of materials is a fascinating blend of both theory and experiment, with experiments pointing the way to useful results in some instances and with theory doing so in others.

材料力学的历史发展过程是一个具有魅力的理论与实验的结合体，在某些情况下是实验指引了通往获取实用结果的道路，而在另一些情况下却是理论。

3. These forces will be continuously distributed over the cross section, analogous to the continuous distribution of hydrostatic pressure over a submerged surface.

这些力将会沿横截面连续分布，类似于作用在被淹没物体表面的连续流体静压分布。

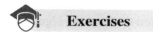
Exercises

Ⅰ. Choose the best answer into the blank.

1. The company is collaborating with partners to develop "____ surfaces" that would allow certain areas of the touch screen to rise up, creating key positions you can feel.
   A. mechanics　　　　B. deformable　　　C. comfortable　　D. smooth
2. ____ Creep is the gradual loss of thickness that may occur if a cushioning material is.
   A. Axial　　　　　　B. transformation　　C. Centroid　　　　D. Compressive
3. America and Russia do not face a problem of the same order of _____ as Japan.
   A. magnitude　　　　B. thought　　　　　C. strain　　　　　D. Compressive
4. His vision is to ____Christianity with "the wisdom of all world religions."
   A. express　　　　　B. think of　　　　　C. convey　　　　　D. blend
5. The bill must clear through the _____ before it becomes legal.
   A. assembly　　　　 B. boss　　　　　　 C. leader　　　　　 D. conference

Ⅱ. Answer the following questions according to the text.

1. What is the aim of the mechanics of materials?
2. What happen to Leonhard Euler?
3. What is needed for the Eq. (2.1)?

Ⅲ. Translate the following into Chinese.

The total elongation of a bar carrying an axial force will be denoted the Greek letter δ, and the elongation per unit length, or strain, is then determined by the equation

$$\varepsilon = \frac{\delta}{L} \tag{2.2}$$

where L is the total length of the bar. Note that the strain ε is nondimensional quantity. It can be obtained accurately from Eq. (2.2) as long as the strain is uniform throughout the length of the bar. If the bar is in tension, the strain is a tensile strain, representing an elongation or a stretching of the material; if the bar is in compression, the strain is a compressive strain, which means that adjacent cross sections of the bar move closer to one another.

Keys

Ⅰ. Choose the best answer into the blank.

1. B   2. D   3. A   4. D   5. A

Ⅱ. Answer the following questions according to the text.

1. Usually the objective of our analysis will be the determination of the stresses, strains, and deformations produced by the loads; if these quantities can be found for all values of load up to the failure load, then we will have obtained a complete picture of the mechanical behavior of the body.

2. By contrast, the famous mathematician Leonhard Euler(1707—1783) developed the mathematical theory of columns and calculated the critical load of a column in 1744, long before any experimental evidence existed to show the significance of his results.

3. A necessary condition for Eq. (2.1) to be valid is that the stress σ must be uniform over the cross section of the bar.

Ⅲ. Translate the following into Chinese.

当杆受轴向力时，其总伸长量用希腊字母 δ 表示。单位长度的伸长量，即应变，可以用如下公式计算得到

$$\varepsilon = \frac{\delta}{L} \tag{2.2}$$

式中：L 为杆的总长度。需要注意的是，应变 ε 是无量纲量。只要应变沿杆的长度方向均匀分布，就可以通过式（2.2）得到其精确结果。如果杆被拉伸，此时的应变称为拉应变，表示材料的伸长量或拉伸量；如果杆被压缩，即为压应变，这就意味着杆相邻截面间的距离变小。

翻译技巧之词量增减（一）

由于历史背景、地理位置、自然环境、民情风俗等方面的巨大差异，英汉两种语言的用词、结构和表达方式不可能完全相同，阅读时应该随时注意两种语言的差别，有时要增补一些词，有时要删减一些词，以符合需要。

词量增补内容如下：

英语专业文章为了避免用词重复，常常省略一些词语，这是阅读时难点之一。还有一些词，在英语中并无含义，但译成中文时应增补，方能正确理解。

有以下几种情况需要增补用词：

（1）增补英语中省略的词。

The best conductor has the least resistance and the poorest the greatest. 良导体的电阻很小。

（2）某些句子结构须增加关联词语：逻辑加词，顺理成章；修辞加词，语气连贯。

<u>Heated</u>, water will change into vapour. 水加热后会变成蒸汽。（增加关联词）

This pointer of the ampere-hour meter moves from zero to two and goes back to zero again. 安时计的指针从零变到二，再变为零。（逻辑加词）

（3）增加表示复数的词。

There is enough coal to meet the world's needs for <u>centuries</u> to come. 煤的储存量能满足世界未来几个世纪的需求。

（4）对具有动作意义的抽象名词增译。

The cost of such a power plant is a relatively small portion of the total cost of the <u>development</u>. 这个电厂的成本相对于发展总成本，只是很小的一部分。

（5）补充概括性的词。

Sino-British links have multiplied—<u>political, commercial, educational, cultural, defense, science and technology.</u> 中英在政治、贸易、教育、文化、防御、科学、技术方面的联系正成倍增加。

Section 2　Introduction of the Structural Mechanics

The objective and task of Structural Mechanics.

Structure is referred to as buildings, constructions or parts of them which can carry loads applied on them and maintain stability and its classification is following:

Table2.1　　　　　　　　　　**Classification of the structure**

Classification	Characteristics	Example
Framed structure	Composed by beams - the main objective of structure mechanics	Beams, arches, rigid frames, trusses
Shells and planes	Thin-walled structure, thickness is much less than length and width	Floors and roofs of buildings
Massive structure	Length, thickness and width has the same order	Gravitational damps in hydraulic structure

The main objective of structural mechanics is framed structures included plane structures that all the members and the applied loads lie in a single plane and otherwise the structures that are spatial structure. This course mainly focuses on plane structures and the demands to structure is strength, stiffness and stability.

The relationship among three mechanics.

(1) Theoretical mechanics mainly studies the basic laws of mechanical motions.

(2) Material mechanics mainly studies the computation methods of typical structural components and the strength of materials, is the introduction to solid mechanics.

(3) Theory of elasticity and plasticity uses the minimal assumptions and uses rigorous analytical methods to study the stress, deformation and fracture of solids, mainly studies shells, plates and massive solids.

Theoretical and material mechanics are the basis and necessary prerequisite for structural mechanics.

Simplified analytical models.

Real structures are complex, it is necessary to simplify the real ones prior to computation, i.e. to replace the real ones by the simplified ones. The simplified representation of sketch of a structure is referred to as its analytical model.

1. The rules of developing simplified model

(1) the analytical model should reflect, as accurately as practically possible, the main stressed and strained characteristics of the structures.

(2) The analytical model should maintain principal factors and eliminate trivial details to simplify the computation.

2. Simplifying points of analytical model

(1) The simplification of structural system: Generally, the actual structures are spatial. But when the parts of structure composed by beams in certain plane carry the loads in the plane, the spatial structure may be divided in several plane structures for analyzing.

(2) The simplification of members: The analytical model is represented by a line diagram in which each member is represented by a line coinciding with its centroidal axis.

3. The simplification of connections

The connections between members are referred to as joints.

(1) Flexible or hinged joint: All member ends connected to a hinged joint have the same translation but may have different rotations.

(2) Rigid joint: The rigid joint prevents relative translations and rotations of the member ends connected to it, i.e. all members connected to it have the same translations and rotations. It can transfer forces and moments.

4. The simplification of the connections between the structure and its foundation

The connections between the structure and its foundation are simplified into supports as follow.

(1) Roller support: idealized as link ,one reaction force.

(2) Hinged support: able to prevent translations in any directions, 2 components of reaction.

(3) Fixed support: preventing the translations and rotations between structure and foundation, 2 reaction forces and moment.

(4) Directional support or double-link support: slides alone its support surface.

5. The simplification of loads

The loads acting on structures are very complex. They can be simplified into concentrated forces and distributed forces.

第二节 结构力学概论

结构力学的研究对象和任务。

结构通常被称为能够承担荷载的建筑物、构筑物，或它们当中的部分的总称，它的分类如下：

表 2.1　　　　　　　　　　　　　结 构 的 分 类

分类名称	特点	实例
杆件结构	由杆件组成，是结构力学的主要研究对象	梁、拱、钢架、桁架
板壳结构	又称薄壁结构，其厚度相比于长度或宽度小得多	建筑物中的楼板和顶板
实体结构	长、宽、厚三个尺寸大小相仿	水工结构中的重力坝

结构力学的主要研究对象是杆件结构，包括所有构件和荷载位于单一平面的平面杆件结构和空间杆件结构。本课程主要侧重于平面杆件结构，其要求包括强度、刚度和稳定性要求。

三种力学之间的关系。

（1）理论力学主要研究机械运动的基本规律。

（2）材料力学是固体力学的学习基础，主要研究典型构件和材料强度的计算方法。

（3）弹塑性力学采用极少的假设与严格的分析方法研究固体的受力、变形与破坏。它主要研究壳、板及实体等结构。

理论力学和材料力学是结构力学学习的基础和必要前提。

简化分析模型。

实际结构往往是很复杂的，因此在进行力学计算之前，有必要将实际结构加以简化，即用一个简化的结构来代替实际结构。这种结构示意图的简化表示通常被称为计算简图。

1. 简图原则

（1）尽可能准确地反映实际结构的主要受力和变形性能。

（2）保留主要特征，略去繁琐的次要因素，使计算简图便于计算。

2. 简化要点

（1）结构体系的简化：一般情况下，实际结构是空间结构。但是，当空间结构的各部分在某一平面内由梁单元组成，且在同一平面上承受负载时，那么可以将该空间结构分解成几个平面结构来进行分析。

（2）杆件的简化：计算简图是一种线图，在其中用纵轴线代替结构中的杆件。

3. 连接部分的简化

结构中杆件相互连接的部分称为结点。

（1）铰接头：与铰接头连接的所有杆件在连接处不能移动，但可转动，即可传递力，但不能传递力矩。

（2）刚结点：刚结点不允许杆件的相对移动和转动，即所有与刚结点相连接的杆件具有相同的移动和转动；也就是说，刚结点既可传递力，又能传递力矩。

4. 结构与其支撑物间连接部分的简化

结构与其支撑物间的连接装置可简化为如下几类支座。

（1）可动铰支座：被支撑结构可绕铰链的铰理想转动；可动铰支座对被支撑结构产生一个过铰且垂直与支撑平面的反力。

（2）铰支座（又称固定铰支座）：被支撑结构不能在铰支座任意方向移动，他们之间只有一对互相垂直的反作用力。

（3）固定支座：被支撑结构相对固定支座（支撑物）既不能移动，又不能转动；固定支座对被支撑结构产生过支撑点的两个相互垂直的反力分量和一个反力矩。

（4）定向支座或平行双链杆座：沿支撑面方向滑动。

5. 荷载的简化

作用在实际结构的荷载通常比较复杂，根据其分布情况可简化为集中荷载和分布荷载等。

 New Words and Expressions

beam	*n.*	横梁
arch	*n.*	拱
truss	*n.*	构架
slide	*n.*	滑动
stiffness	*n.*	硬度
motion	*n.*	运动
elasticity	*n.*	弹性
plasticity	*n.*	塑性
minimal	*adj.*	最小的
rigorous	*adj.*	严格的
shell	*n.*	壳
prerequisite	*n.*	先决条件
computation	*n.*	估算
hinged joint		铰接头
rigid joint		刚性结点
rigid frame		刚性构架
hydraulic structure		水工结构

 Notes

1. Theory of elasticity and plasticity uses the minimal assumptions and uses rigorous analytical methods to study the stress, deformation and fracture of solids, mainly studies shells, plates and massive solids.

弹塑性力学采用极少的假设与严格的分析方法研究固体的受力、变形与破坏。它主要研究壳、板及实体等结构。

2. But when the parts of structure composed by beams in certain plane carry the loads in the plane, the spatial structure may be divided in several plane structures for analyzing.

但是，当空间结构的各部分在某一平面内由梁单元组成，且在同一平面上承受负载时，那么可以将该空间结构分解成几个平面结构来进行分析。

Exercises

Ⅰ. Choose the best answer into the blank.

1. If you eat soon after your workout, yon can minimize muscle____and soreness.
   A. uncomfortable   B. uneasy   C. stiffness   D. healthy
2. These are____standards, though-all they do is open a channel.
   A. easiest   B. minimal   C. biggest   D. needed
3. Our teacher is so____that he seldom lets up on us.
   A. good   B. handsome   C. kind   D. rigorous
4. In fact, if you like, you can check names by regular expression or by other____.
   A. computation   B. fact   C. way   D. tool
5. A command of information is the necessary _____ to the scientific consideration of any subject.
   A. process   B. thing   C. step   D. prerequisite

Ⅱ. Answer the following questions according to the text.

1. What do shells and planes include?
2. What is the strictest mechanics among three in the essay?
3. How can we simplify the connections?

Ⅲ. Translate the following into Chinese.

The simplification of material property:

In civil engineering the structures are made of materials such as steel, concrete, bricks, stone, timber and so on, to simplify the analysis all structural materials are assumed to be continuous, homogeneous, isotropic and perfectly elastic.

Above assumptions are suitable for metal within some stressing extent, but for concrete, reinforced concrete, bricks, stone and the like, the assumptions will have some degree of approximation. As far as timber, because the property alone timber grain is quite different from that cross the timber grain, the attention should be paid when applying the assumption.

Keys

Ⅰ. Choose the best answer into the blank.
1. C   2. B   3. D   4. A   5. D

Ⅱ. Answer the following questions according to the text.
1. Floors and roofs of buildings.
2. structural mechanics.

3. The connections between members are referred to as joints.

1) Flexible or hinged joint: All member ends connected to a hinged joint have the same translation but may have different rotations.

2) Rigid joint: The rigid joint prevents relative translations and rotations of the member ends connected to it, i.e. all members connected to it have the same translations and rotations. It can transfer forces and moments.

Ⅲ. Translate the following into Chinese.

材料特性的简化：

在土木工程中，结构通常由钢、混凝土、砖、石、木料等材料做成。为了简化分析，假设所有结构的材料为连续、均匀、各向同性、完全弹性或弹塑性的。

上述假设对于金属材料在一定受力范围内是适用的，但是对于混凝土、钢筋混凝土、砖、石等材料则带有一定程度的近似。至于木材，因其顺纹与横纹方向物理性质的不同，故应用这些假设时应注意。

翻译技巧之词量增减（二）

词量删减如下：

英语中常用一些关系词、冠词、连词、代词、同义词及同位语等，在译成汉语时可以免去不译。

有以下几种情况需要省略用词：

1. 冠词的省译

With an equation we can work with an unknown quantity. 利用方程，我们可以求解未知量。

2. 介词的省译

Hydrogen is the lightest element with an atomic weight of 1.008. 氢是最轻的元素，其原子的质量为 1.008。

3. 人称代词、物主代词、反身代词的省译

When the signal we pick up has increased by 10 times as the gain may have been reduced by 8 times. 信号增大到 10 倍，增益降低到 1/8。

4. "it" 在某些句型（如强调、形式主语、形式宾语）中的省译

It is said that numerical control is the operation of machine tools by numbers. 人们说，数控就是机床用数字加以操纵。

5. 连词的省译

Using a transformer, power at low voltage can be transformed into power at high voltage. 如果使用变压器，低电压的电能就能转换成高电压的电能。

6. 关系代词的省译

An obvious disadvantage of a long air gap is that it necessitates a greater field MMF to produce a specified air-gap flux. 长空气隙一个明显的缺点就是需要较大的磁场磁动势以产生规定的气隙磁通量。

Section 3　Unbalanced Tension Analysis for UHV Transmission Towers in Heavy Icing Areas

Unbalanced tension is one of the most important controlling loads for the design of transmission towers in cold regions. It can be caused in some cases including non-uniform accreted ice, broken wires, ice shedding and galloping. Firstly, for the comparison of the unbalanced tensions and the inclinations of the suspension strings, the FEA method and the constant conductor length method were applied. The results by two methods are approximately well agreed. With the consideration of some design parameters of the transmission line, a seven continuous span FEA model of conductors and insulators was established. Secondly, under different load cases, the tensions and the unbalanced tensions of conductors were analyzed for the UHV suspension tower and tension tower in heavy icing areas. It shows that the load modes and the eccentricity of the accreted ice, as well as the wind velocity only have little effect on the unbalanced tensions. The transforming density method and 10m/s of wind velocity are proposed for the analysis of the unbalanced tensions.

When there are no elevation difference and span difference, the unbalanced tension of the UHV suspension tower increases with the ice thickness, the span length and the icing rate, and the calculated values for different ice thickness are lower than those of regulations. With the increasing of the elevation difference and the span difference, the ratios of the unbalanced tensions with and without elevation difference as well as span difference increase. However, the ratios of the unbalanced tensions with and without elevation difference as well as span difference decrease with the ice thickness. For tension towers, variations of the elevation difference and the span difference have little effect on the unbalanced tensions. Lastly, the calculated values of the unbalanced tension percentages of UHV transmission towers were compared with those of applicable regulations, some suggestions on the unbalanced tension values were proposed.

As a normal type of load in icing areas, the non-uniform accreted ice may induce longitudinal unbalanced tensions in two adjacent spans of the suspension tower or tension tower. The suspension tower will be subjected to bending moment or torsional moment, which is the main reason for damage or collapse of suspension towers. After the collapse of a suspension tower, suspension towers at the two adjacent sides will bear much higher unbalanced tensions and impact loads, and the cascades in transmission lines may be induced.

In recent years, some researchers and designers have been focused on the study of the unbalanced tensions under non-uniform accreted ice. Liao et al. (2006) computed the unbalanced tensions for the 1000kV UHV suspension tower in light icing areas. For UHV suspension towers in plain areas, hill areas and mountain areas, some advices for the determination of the unbalanced tension values were proposed. In the report by China Power Engineering Consulting Group Cooperation (2008), a seven continuous span conductor model was established. Unbalanced tensions for the 220~750kV suspension towers and tension towers were calculated with ice thickness of 10mm to 50mm. Based on the constant conductor length method, Zhang and Liu (2009) developed a program for the analysis of the unbalanced tensions in continuous spans. Unbalanced tensions and inclinations of suspension strings were calculated for suspension towers in light icing areas. It is concluded that the unbalanced tension values calculated by the Design Code of 110~750kV Overhead Transmission Lines (National Energy Administration, 2010) are relatively low for some cases, and the unbalanced tension values should be enhanced for suspension towers in the transmission lines not higher than 500kV. Cheng and Xue (2011) studied the effects of some design parameters on the unbalanced tensions. The design parameters include safety coefficient of conductors, length of suspension strings, span numbers, span difference and elevation difference.

After the serious ice disaster in South China in 2008, in order to ensure the anti-bending and anti-torsion capacity of transmission towers, the Technical Code for Designing of Overhead Transmission Line in Medium & Heavy Icing Area (China Electricity Council, 2009) was revised, and the unbalanced tension percentages and icing rates in two adjacent sides of transmission towers were regulated. The ratio of the unbalanced tension to the maximum working tension of conductors is defined as the unbalanced tension percentage. This code is suitable for 110~750kV transmission lines, however regulations or suggestions on the unbalanced tension values for UHV transmission towers were not mentioned.

For the calculation of unbalanced tension in IEC 60826 (International Electrotechnical Commission, 2003), the two adjacent spans are loaded with accreted ice of $0.7g_R$ and $0.28g_R$ is the reference ice weight accreted on the conductors. That means the icing rates of the two adjacent spans are constant values. However, in the Technical Code for Designing of Overhead Transmission Line in Medium & Heavy Icing Area (China Electricity Council, 2009), the icing rates are varied for different transmission line classes or different types of towers.

With the development of the UHV transmission projects, the UHV transmission lines cannot be avoided to pass the heavy icing areas. The XiangJiaBa-ShangHai ±800 kV DC transmission line in operation has passed heavy icing areas in Southwest China. The YaAn-NanJing 1000kV AC transmission line in consideration will pass the icing areas with 20mm accreted ice in Southwest China. According to the design conditions of the UHV transmission lines, it is noted that the ice thickness referred in this paper is the design ice thickness with a return period of 100 years. Until recently, few studies have dealt with the unbalanced tensions for the UHV transmission towers in heavy icing areas.

Numerical simulations using nonlinear finite element analysis (FEA) are useful to study the unbalanced tensions for UHV transmission towers. A seven continuous span conductor-string model in UHV transmission line was developed. Some design parameters were considered in the FEA model, which included the loading mode of accreted ice, the eccentricity of accreted ice, the wind velocity, the ice thickness, the icing rate, the span length, the elevation difference and the span difference. Parametric study on the unbalanced tensions and the unbalanced tension percentages was performed. Based on the analysis results, the unbalanced tension values were determined for the UHV transmission towers in heavy icing areas. This study should be useful for ensuring the safety and reliability of UHV transmission towers in heavy icing areas.

 原文翻译

第三节　重覆冰区特高压输电杆塔不平衡张力分析

在寒冷地区的输电塔设计中，不平衡张力是最重要的控制荷载参数之一。产生不平衡张力的原因包括不均匀覆冰、断线、脱冰跳跃和舞动等。首先，应用有限元分析方法和等线长法比较了不平衡张力与悬垂串偏移量之间的关系，发现这两种方法的计算结果基本一致。为了考虑多种线路设计参数的影响，建立了连续 7 挡导线—绝缘子有限元模型。其次，在不同的负载情况下，分析了重覆冰区特高压直线杆塔和耐张杆塔的导线张力及不平衡张力。结果表明，覆冰加载模式、覆冰偏心和风速对不平衡张力影响不大，建议采用换算密度法和 10m/s 风速来分析不平衡张力。

当无高差和挡距差时，随着覆冰厚度、挡距以及覆冰率的增加，特高压直线杆塔的不平衡张力逐渐增加，且不同覆冰厚度下的不平衡张力计算值均小于规程规定值。随着高差和挡距差的增加，有高差和挡距差的不平衡张力与无高差和挡距差的不平衡张力比值增大；但是，随着冰厚的增加，不平衡张力比值减小。对于耐张塔而言，高差和挡距差的变化对不平衡张力的影响很小。最后，将特高压输电塔不平衡张力百分数的计算值与规程规定值做了比较，并对不平衡张力值提出了一些建议。

作为覆冰区的一种典型荷载，不均匀覆冰可能会导致直线塔或耐张塔出现纵向不平衡张力，从而使直线塔受到弯矩或扭矩的作用，这是直线塔损坏或垮塌的主要原因。直线塔垮塌后，会导致旁边两侧的直线塔承受更大的不平衡张力和冲击力，因此往往会引发串倒，扩大事故范围。

近年来，不均匀覆冰工况下的杆塔不平衡张力研究一直受到科研和设计人员的关注。廖等人在 2006 年通过计算轻覆冰区 1000kV 特高压直线塔的不平衡张力，给出了平地、丘陵及山地区特高压直线塔不平衡张力的取值建议。中国电力工程顾问集团公司在 2008 年采用连续 7 挡模型计算了 220～750kV 线路、10～50mm 冰厚直线塔和耐张塔的不平衡张力。在 2009 年，张和刘基于等线长法编制了连续挡不平衡张力的计算程序，并计算了轻覆冰区直线塔的不平衡张力及绝缘子偏移量，认为在一些情况下《110kV～750kV 架空输电线路设计规范》(国家能源局，2010) 关于导线不平衡张力的取值偏小，同时对于 500kV 以下电压等级线路的直线塔，导线不平衡张力的取值应适当提高。在 2011 年，程和薛分析了导线安全系数、绝缘子串长、挡距数、挡距差和高差等设计参数对覆冰不平衡张力的影响。

我国南部经历了 2008 年严重冰灾之后，为了保证杆塔的抗弯和抗扭能力，修订了《中重覆冰区架空输电线路设计技术规程》(中国电力企业联合会，2009年)，同时调整了输电杆塔相邻两侧的不平衡张力百分数和覆冰率。其中，导线的不平衡张力与其可承受最大张力的比值称为不平衡张力百分数。该技术规范适用于 110～750kV 输电线路，然而，尚未有特高压输电杆塔不平衡张力值的规程或建议。

按照 IEC 60826（国际电工委员会，2003）不平衡张力计算规定，相邻两挡导线的参考覆冰重量为 $0.7g_R$（g_R 是一种重量单位，格令，最初在英格兰定义一颗大麦粒的重量为 1 格令）、$0.28g_R$，所以这意味着其将相邻两挡的覆冰率看作定值。然而，在《中重覆冰区架空输电线路设计技术规程》(中国电力企业联合会，2009)中，覆冰率是随着输电线路的类别或铁塔的不同类型而变化的。

随着特高压输电工程的发展建设，特高压输电线路不可避免地要通过重覆冰区。已经投运的±800kV 向家坝—上海直流输电线路经过了我国西南部的重覆冰区；拟建的 1000kV 雅安—南京交流输电线路将经过我国西南部覆冰厚度为 20mm 的重覆冰区。需要指出的是，根据特高压输电线路的设计条件，本文中提及的覆冰厚度是设计覆冰厚度（修订周期为 100 年）。目前，还没有针对重覆冰区特高压输电铁塔不平衡张力的研究成果。

基于非线性有限元分析的数值模拟方法，对于特高压输电杆塔不平衡张力的研究具有重要作用。通过重冰区特高压输电线路连续 7 挡导线—绝缘子有限元分析模型，考虑覆冰加载模式、覆冰偏心、风速、冰厚、覆冰率、挡距、高差以及挡距差等因素的影响，从而研究不同参数对不平衡张力和不平衡张力百分比的影响规律。根据分析结果，确定了重覆冰区特高压输电线路杆塔的不平衡张力值。本文的研究成果对于确保重覆冰区特高压输电线路杆塔的安全及稳定性具有重要意义。

New Words and Expressions

longitudinal	*adj.* 纵向的，长度的
eccentricity	*n.* 偏心率
adjacent	*adj.* 相邻的
span	*n.* 跨度，挡距
parameter	*n.* 参数，系数
collapse	*n.* 倒塌，失败
UHV	超高压
unbalanced tension	不平衡张力
ice shedding	冰脱落
ice galloping	冰舞动
suspension strings	悬垂绝缘子串
tension towers	耐张杆塔
impact load	冲击负载
transmission lines	输电线路
safety coefficient	安全系数
anti-bending	抗弯

anti-torsion	抗扭
numerical simulation	数值仿真
finite element analysis	有限元分析

Notes

1. When there are no elevation difference and span difference, the unbalanced tension of the UHV suspension tower increases with the ice thickness, the span length and the icing rate, and the calculated values for different ice thickness are lower than those of regulations.

当无高差和挡距差时，随着覆冰厚度、挡距以及覆冰率的增加，特高压直线杆塔的不平衡张力逐渐增加，且不同覆冰厚度下的不平衡张力计算值均小于规程规定值。

2. As a normal type of load in icing areas, the non-uniform accreted ice may induce longitudinal unbalanced tensions in two adjacent spans of the suspension tower or tension tower. The suspension tower will be subjected to bending moment or torsional moment, which is the main reason for damage or collapse of suspension towers.

作为覆冰区的一种典型荷载，不均匀覆冰可能会导致直线塔或耐张塔出现纵向不平衡张力，从而使直线塔受到弯矩或扭矩的作用，这是直线塔损坏或垮塌的主要原因。

3. That means the icing rates of the two adjacent spans are constant values.

这意味着其将相邻两挡的覆冰率看作定值。

Exercises

Ⅰ. Choose the best answer into the blank.

1. Under different ____ cases, the tensions and the unbalanced tensions of conductors were analyzed for the UHV suspension tower and tension tower in heavy icing areas.
   A. load　　　　　　B. voltage　　　　　C. current　　　　　D. resistance

2. Unbalanced tension is one of the most important controlling loads for the design of ____ towers in cold regions.
   A. tension　　　　　B. transmission　　　C. angle　　　　　　D. suspension

3. With the increasing of the ____ difference and the span difference, the ratios of the unbalanced tensions with and without elevation difference as well as span difference increase.
   A. elevation　　　　B. span　　　　　　C. parameter　　　　D. eccentricity

4. The design parameters include safety ____ of conductors, length of suspension strings, span numbers, span difference and elevation difference.
   A. coefficient　　　B. collapse　　　　　C. parameter　　　　D. span

5. Numerical ____ using nonlinear finite element analysis (FEA) are useful to study the unbalanced tensions for UHV transmission towers.
   A. analysis　　　　B. simulations　　　C. collapses　　　　D. coefficients

Ⅱ. Answer the following questions according to the text.

1. What circumstance will create unbalanced tensions for UHV transmission towers in heavy icing areas?

2. What method is used to study the unbalanced tensions for UHV transmission towers?

Ⅲ. Translate the following into chinese.

Based on the analysis results, the unbalanced tension values were determined for the UHV transmission towers in heavy icing areas. This study should be useful for ensuring the safety and reliability of UHV transmission towers in heavy icing areas.

Keys

Ⅰ. Choose the best answer into the blank.

1. A　　2. B　　3. A　　4. A　　5. B

Ⅱ. Answer the following questions according to the text.

1. It can be caused in some cases including non-uniform accreted ice, broken wires, ice shedding and galloping.

2. the FEA method and the constant conductor length method.

Ⅲ. Translate the following into chinese.

根据分析结果，确定了重覆冰区特高压输电线路杆塔的不平衡张力值。本文的研究成果对于确保重覆冰区特高压输电线路杆塔的安全及稳定性具有重要意义。

翻译技巧之语序的调整

语序调整就是根据两种语言的不同语序特点对句子成分进行调整安排。两种语言在时间、地点、逻辑等表述顺序上常常存在一些差异。英译汉时，为了使译文合乎汉语语法和语言习惯，需要改变语序。英语很多后置的定语、状语和修饰语，在翻译中的位置变化是英汉翻译中最常见也是最基本的现象。如：

1. 定语位置的调整

英语定语就其位置来说，可以分为前置定语和后置定语，而汉语定语一般放在名词之前，放在名词之后的极少，因此，英译汉时需要注意定语位置的调整。如：

The house whose roof was damaged has now been repaired. 屋顶坏了的房子现在已经修好了。

2. 状语顺序的调整

在汉语里状语多数放在所说明或修饰的词的前面，但英语状语有的放在句子最后，有的放在句中，有的放在句首，翻译时要考虑汉语表述习惯，进行语序调整。如：

Many types of animals have now vanished from the earth. 许多动物在地球上灭绝了。

3. 同位语、插入语成分语序的调整

插入语和同位语一般是对一个事物或一句话做一些附加的解释，英语中插入语和同位语位置十分灵活，译成汉语时，有时要做相应的调整以合乎汉语的表达习惯。如：

There are, to be exact, only two choices. 准确地说，只有两种选择。

4. 评论和表态中语序的调整

英语的语序和汉语不同，英语先评论或先表态，再说有关情况，而汉语则相反，翻译时要按汉语习惯重新组织和排列句子的语序。如：

Food is essential to life. 食物是维持生命不可或缺的。

It is useless to complain all the time. 总抱怨是没用的。

It is unlucky to walk under a ladder in the western countries. 在西方，人走在梯子下面是不吉利的。

5. it 结构的语序调整

在英语里，主语可以后置，前面用形式主语 it 代替。在汉语里，主语放在谓语的前面，不能用代词代替。在翻译"It +系动词+实际主语"的结构时，主语应尽可能提前，而且应该明确具体，要将做真正主语的不定式等连同后面的从句提前译在主语的位置上。如：

It has long been proved that the creative power of the people knows no limits. 人民的创造力是无穷的，这一点早已得到证实。

Chapter 3   Corona Discharge

Section 1   Comparison of Methods for Determining Corona Inception Voltages of Transmission Line Conductors

1   Introduction

Corona discharge phenomena are quite common on both AC and DC transmission lines when the surface electric field is sufficiently high. Corona results in audible noise (AN), radio frequency interference (RFI) and corona losses.

Much research has been done in recent years in this area. Vinh statistically analyzed long-term audible noise and corona loss data under a variety of weather conditions and Chartier investigated RFI and AN levels from particular conductors, but neither compared the inception voltages found from these different phenomena or the problem of defining the actual inception voltages under these conditions.

The corona inception voltage on cylinders was investigated by Peek who thereby developed his well-known empirical equation. Successful attempts have been made to relate the corona inception voltage to the physical mechanisms of discharge with various electrodes and gases, for AC and DC and various environmental factors. Yamazaki and Amoruso developed calculations of the effect of stranding on the corona inception voltage but did not make measurements on actual conductors. More recently UV corona imagers have been used to detect corona under outdoor daylight conditions.

In the present work a corona cage was used. This is a single-phase test facility in which conductors (single or bundled) are centered in a grounded cage. The close proximity of the grounded cage allows the surface electric field to be at the levels found on transmission lines in the field but at lower applied voltages. Corona cages have therefore been widely applied for predicting the corona characteristics of transmission lines.

Two new transmission-line conductors, of different construction, were used with a corona cage in the work described here. They were single conductors, not bundled, with a steel (7-strand) core surrounded by several layers of aluminum stranding. The behavior below, during and after the onset of corona was observed via four measurement systems: an ultraviolet (UV) imager, a partial discharge (PD) detector, a sound level meter and an RFI receiver (measuring the high-frequency component of the corona current).

A method to determine the mean inception voltage was devised and applied to the output of the above instruments. The outputs and these inception voltages were compared with each other and the results of calculations using Peek's equation and a direct 'firstprinciples' approach.

2 Corona cage and conductors

2.1 Corona cage

The small corona cage designed at Tsinghua University and showed in Fig.31 was used in this work. The dimensions of the inner cage are a square cross-section of 1.7m^2 and a length of 4m, comprising a 3m long central section for measurements and two 0.5m-long guard sections to eliminate the end effects.

2.2 Experiment equipment

The two conductors employed in the present work were standard transmission-line conductors manufactured in China, having the codes LGJ 500/35 and LGJ 400/50. The former (hereinafter Conductor A) has 7 steel strands surrounded by 45 aluminum strands, the latter (hereinafter Conductor B) has 7 steel strands surrounded by 54 aluminum strands. The cross-sectional areas of the steel and aluminum are approximately 35 and 500mm^2 respectively (Conductor A); and 50 and 400mm^2 (Conductor B) ehence the 500/35 and 400/50 appellations.

Fig.3.1 General views of the corona cage used in the experiment

UV, PD, current and audible noise (AN) levels were measured as indicated in Fig3.2. The HV AC source was set as 50Hz. These experiments were carried out in Wuhan, Hubei Province, PRC, where the altitude is 23 m and the summer temperature was in the range 36~40℃, with the relatively humidity in the range 47%~59%.

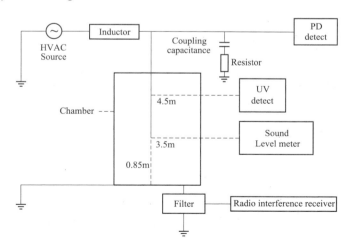

Fig.3.2 Schematic diagram of the measurement system and corona cage.

3 The four instruments and related experimental results

3.1 Measurement of UV photon output

An Ofil Corporation Superb ultraviolet imager was used to measure the photon production by electric discharges at irregularities on the conductor surface. Such UV imagers have been widely used for corona discharge detection in power systems and are discussed in greater detail.

Detector was placed in the same height as the conductor and with a sight-line normal to the conductor.

Fig3.3 shows three sets of measurements for Conductor A under virtually the same atmosphere conditions, and shows good repeatability. When the applied voltage was less than 100kV, there was virtually no UV photon count so no corona discharge was occurring. As the voltage was increased, the corona discharge points grew in number and the photon detection began to increase gradually and then very rapidly above about 125kV. Thus there are 4 regions: the no-corona region where the graph is flat and with zero gradient; the corona region where the graph has increasing gradient and strong positive curvature; the straight line region where the line has a high gradient and zero curvature; and finally there is often a small region where the gradient increases slightly (as in this case) or decreases slightly (as in some of the cases seen later). Using these definitions, the steep, zero-curvature third region could be determined and the best straight line through it found by regression analysis (least-squares method). This line was extended to intersect with the extended line through the first region data points. The voltage at this point was defined as the inception voltage and was found to be 123.4kV. Using the results for conductor B, which are plotted as in Fig3.4, the inception voltage was found to be 118.3kV.

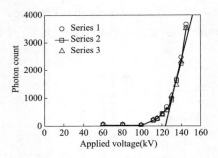

Fig3.3　UV photon count versus applied voltage for conduct A.

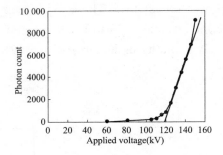

Fig3.4　UV photon count versus voltage for conduct B.

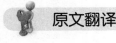

第一节　输电线路电晕起始电压确定方法的比较

1　前言

电晕现象是由于导体表面电场较高产生的,并且在交、直流输电线路中非常常见。电晕会造成可听噪声、无线电干扰及电晕损耗等不良影响。

近年来针对电晕问题做了大量相关研究。容统计地分析了在各种气候条件下长期的可听噪声值和电晕损耗值,而沙尔捷研究了特殊导体的无线电干扰和可听噪声水平,但这些研究都未能发现这些特殊情况下电晕起始电压的不同，也未能确定这些特殊情况下的实际起始电压。

皮克研究了圆柱体的起始电晕电压，进而建立了著名的经验公式。在他的研究中，尝试将电晕起始电压与各种电极和气体放电的物理机理相联系，该研究在交、直流电等不同的环境因素下都取得了成功。山崎和阿莫鲁索在电晕起始电压的计算中，对导体线模型起始电压进行了考虑，但他们并没有对实际导体进行实验检测。最近，越来越多的紫外线电晕成像技术在白天户外的环境下被用于检测电晕。

在最近的研究中经常使用的实验方法是电晕笼实验。这是个单相的测试装置，导体（单股或者多股）放在一个接地的金属笼中间。在这个接地笼实验中的导体表面允许电场和电压较低的输电线路电场处在同一个水平上。因此，电晕笼实验被广泛应用于输电线路电晕预测中。

两个不同构造的输电线路导线在电晕笼中进行上述实验。这些导线不是多股的，而是单股的，由钢芯（7 股）与外层包裹的铝绞线构成。在电晕发生的各个过程将通过四种不同的测量仪器进行观测：紫外线（UV）成像仪、局部放电（PD）检测仪、噪声计和射频干扰接收器（测量电晕电流中的高频部分）。

以上仪器会对测量的平均起始电压值进行计算，这是由这些仪器的设计所决定的。最终的输出结果将与起始电压平均值，运用 Peek 公式所计算的结果，以及直接使用"第一原理"计算的结果三者进行比较而确定。

2 电晕笼和导线

2.1 电晕笼

图 3.1 是清华大学所设计的小型电晕笼，这个电晕笼用于电晕特性的相关试验中。该电晕笼横截面积为 $1.7m^2$，长为 4m，以及一个长为 3m 用于测量的中心部分和两个 0.5m 长的消除边缘效应的防护单元所组成。

2.2 实验装置

当前试验中所使用的导线是中国制造的标准输电线路导线，其型号分别为 LGJ500/35 和 LGJ400/50。前者（下文中的导线 A）是由 7 根钢芯和 45 根铝导线组成，后者（下文中的导线 B）是由 7 根钢芯和 54 根铝导线组成。导线 A 的钢芯和铝导线的横截面面积分别为 $35mm^2$ 和 $500mm^2$；导线 B 的横截面面积分别是 $50mm^2$ 和 $400mm^2$，这就是 LGJ500/35 和 LGJ400/50 所表示含义。

图 3.1 试验中使用的电晕笼总体图

紫外线、局部放电、电流和可听噪声分别按照图 3.2 所示方式进行测量。高压交流电源的频率是 50Hz。这些试验在湖北武汉进行，当地海拔是 23m，温度是 36～40℃，相对湿度为 47%～59%。

3 四种仪器及其试验结果

3.1 紫外线的测量

Ofil 公司的高级紫外线成像仪将用来测量不规则导体表面的电流所产生的光子。这种紫外线成像仪被广泛地用于电力系统中电晕的检测与更严谨的学术研究中。

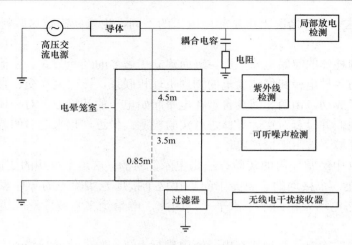

图 3.2 电晕笼和检测系统流程图

检测仪被放在与导线相同的高度上，因此，导线可被等效成一根单股导线。

在几乎相同的大气条件下，图 3.3 显示的是导线 A 的三组测量数据，并且展示了非常好的可重复性。当外加电压低于 100kV 时，几乎没有紫外线光子，同时也没有发生电晕现象。随着电压的升高，电晕放电点也在升高，光子的测量数据也开始呈现逐渐增长的趋势，当电压达到 125kV 及以上时，增速更加明显。然而，整体图像可分为 4 个明显区域：在图像平缓没有梯度的地方是未发生电晕的区域；梯度逐渐增长并且迅速弯曲的地方是电晕发生区域；有高梯度但没有弯曲的地方是直线区域；最后还有一个区域，这个区域里的梯度增长很慢（像这个例子一样）或者减少很慢（和下文中将出现的例子一样）。根据这些定义，高梯度、零弯曲的第三个区域能被判定，而且可通过线性回归分析（最小二乘法）获得最接近的拟合直线。这条直线延伸出来和第一个区域延伸出来的线相交。这一点被定义为起始电压点，它的值是 123.4kV。如图 3.4 所示，对于导体 B 也使用同样的方法判断，得到的起始电压为 118.3kV。

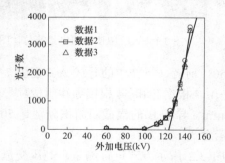 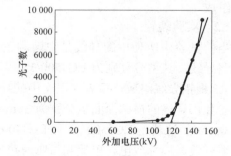

图 3.3 导体 A 的光子计数和外加电压图　　图 3.4 导体 B 的光子计数和外加电压图

 New Words and Expressions

radiofrequency	*n.* 高频；高周波
statistically	*adj.* 系统地
electrodes	*n.* 电极；电焊条

single	*adj.* 单股的
corona	*n.* 电晕
grounded	*adj.* 接地的
HV	*n.* 高压
source	*n.* 电源
a charge simulation method	模拟电荷法
the electric field	电场
corona inception voltage	电晕起始电压
partial discharge detector	局部放电检测仪
corona cage	电晕笼

 Notes

1. Vinh statistically analyzed long-term audible noise and corona loss data under a variety of weather conditions and Chartier investigated RFI and AN levels from particular conductors, but neither compared the inception voltages found from these different phenomena or the problem of defining the actual inception voltages under these conditions.

容系统地分析了在各种气候条件下长期的可听噪声值和电晕损耗值，而沙尔捷研究了特殊导体的无线电干扰和可听噪声水平，但这些研究都未能对比发现这些不同现象中的电晕的起始电压的不同，与此同时也未确定这些特殊情况下的实际起始电压。

2. The close proximity of the grounded cage allows the surface electric field to be at the levels found on transmission lines in the field but at lower applied voltages.

在这个接地笼实验中的导体表面允许电场和电压较低的输电线路电场处在同一个水平上。因此，电晕笼实验被广泛应用于输电线路电晕预测中。

3. The outputs and these inception voltages were compared with each other and the results of calculations using Peek's equation and a direct "firstprinciples" approach.

最终的输出结果将与起始电压平均值，运用 Peek 公式所计算结果，以及直接使用"第一原理"计算的结果三者进行比较而确定。

 Exercises

I. Choose the best answer into the blank.
1. ____ times current is equal to voltage.
   A. Resistance   B. Inductance   C. Capacitance   D. Conductance
2. The length of circle is equal to double π and ____ .
   A. diameter   B. area   C. radius   D. proportion
3. Which of the following does not belong to the metal? ____ .
   A. steel   B. aluminum   C. copper   D. rubber
4. We're interested in the source of these ____ rumors.

    A. true   B. fictitious   C. interested   D. false

5. This can reduce the ____ of investment, time and labor.

    A. double   B. produce   C. use   D. duplication

II. Answer the following questions according to the text.

1. What is the Peek equation?

2. Why do the corona cage choose the square cross-section?

3. What is the meaning of the LGJ500/35?

III. Translate the following into Chinese.

    Comparing the data and curves produced by the UV imager, the PD detector, the AN meter and the RFI receiver, it is seen that closely similar results (within a roughly 2% range) were obtained for the inception voltage as defined by the downward extension of a trend line through the straight portion of the graph after the curved transition region.

    It should be noted that the four measuring systems measure slightly different amounts of the corona: the UV imager only detects UV from the discharges on the conductor surface facing the instrument; the PD detector detects discharges over the whole of the conductor surface; the AN meter detects sound from the discharges on the conductor surface facing it rather better than those at the back; and the RFI receiver, like the PD detector, detects the current from discharges over the whole of the conductor surface.

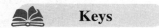

Keys

I. Choose the best answer into the blank.

1. A   2. C   3. D   4. B   5. D

II. Answer the following questions according to the text.

1. The corona inception voltages of cylindrical conductors were investigated experimentally many years ago by Peek, who formulated empirical equations for the inception field and voltages.

2. In the present research, a square cross-section corona cage was used for ease of manufacture.

3. The cross-sectional areas of the steel and aluminum are approximately 35 and 500mm^2 respectively.

4. There are four methods to determine the inception voltage: UV photon, radio interference, audible sound, partial discharge, Peek equation.

5. 201.1kV.

III. Translate the following into Chinese.

    比较紫外线成像仪、局部放电检测仪器、可听噪声计和无线电干扰检测仪的数据和图像，我们可以知道他们有类似的结果（误差范围在2%以内），这些获取的起始电压通过图形圆弧的切线向下延长得到的。

    应该注意的是这4种测量的方法测量的电晕有一点轻微的不同：紫外线成像仪仅仅检测面对仪器的导体表面电流；局部放电检测仪检测的是整个的导体表面放电；可听噪声计检测的声音是面对仪器的放电的导体表面而不是背部；和局部放电检测仪类似，无线电干扰接收器检测的是整个导体表面的放电电流。

翻译技巧之语序的调整

语序调整就是根据两种语言的不同语序特点对句子成分进行调整安排。两种语言在时间、地点、逻辑等表述顺序上常常存在一些差异。英译汉时，为了使译文合乎汉语语法和语言习惯，需要改变语序。英语很多后置的定语、状语和修饰语，在翻译中的位置变化是英汉翻译中最常见也是最基本的现象。

1. 定语位置的调整

英语定语就其位置来说，可以分为前置定语和后置定语，而汉语定语一般放在名词之前，放在名词之后的极少，因此，英译汉时需要注意定语位置的调整。如：

The house whose roof was damaged has now been repaired. 屋顶坏了的房子现在已经修好了。

2. 状语顺序的调整

在汉语里状语多数放在所说明或修饰的词的前面，但英语状语有的放在句子最后，有的放在句中，有的放在句首，翻译时要考虑汉语表述习惯，进行语序调整。如：

Many types of animals have now vanished from the earth. 许多动物在地球上灭绝了。

3. 同位语、插入语成分语序的调整

插入语和同位语一般是对一个事物或一句话做一些附加的解释，英语中插入语和同位语位置十分灵活，译成汉语时，有时要做相应的调整以合乎汉语的表达习惯。如：

There are, to be exact, only two choices. 准确地说，只有两种选择。

4. 评论和表态中语序的调整

英语的语序和汉语不同，英语先评论或先表态，再说有关情况，而汉语则相反，翻译时要按汉语习惯重新组织和排列句子的语序。如：

Food is essential to life. 食物是维持生命不可或缺的。

It is useless to complain all the time. 总抱怨是没用的。

It is unlucky to walk under a ladder in the western countries. 在西方，人走在梯子下面是不吉利的。

5. it 结构的语序调整

在英语里，主语可以后置，前面用形式主语 it 代替。在汉语里，主语放在谓语的前面，不能用代词代替。在翻译"It＋系动词＋实际主语"的结构时，主语应尽可能提前，而且应该明确具体，要将做真正主语的不定式等连同后面的从句提前译在主语的位置上。如：

It has long been proved that the creative power of the people knows no limits. 人民的创造力是无穷的，这一点早已得到证实。

Section 2 Air–Gap Discharge Characteristics in Foggy Conditions Relevant to Lightning Shielding of Transmission Lines

1. Introduction

In designing large-scale transmission lines, outages caused by lightning are a significant problem. Electrogeometric models have been used in respect of the shielding of transmission lines. Young et al. first introduced the concept of the electrogeometric model using the relationship between the potential of the downward leader channel and return stroke current magnitude. In 1968, Armstrong and Whitehead improved the electrogeometric model, taking the stroke angle into consideration, proposing the formula of striking distance R_s in a relatively simple form as shown

$$R_s = a \times I^b \tag{3.1}$$

Where I is the lightning peak current magnitude and a and b are constants.

The constants a and b are not theoretically determined and the validity of these values has been verified only by comparison between the calculated outage rates of transmission lines using the model and field experiences. In fact, different values for the constants a and b have been proposed in related references. Nevertheless, the model proposed by Armstrong and Whitehead (A-W model) is widely used for the lightning protection design of transmission lines.

However, after ultra-high voltage (UHV)-designed transmission lines began operation at 500 kV in Japan, more lightning faults occurred than had been calculated using the A-W model. Results, based on observations of the distribution of direct lightning strokes to phase conductors, were quite different than those predicted by the A-W model in the design stage.

Recently, more sophisticated shielding models have been proposed in related references. These propose that the striking distance is not only a function of the return stroke current magnitude but also the height of the earthed object. According to these models, the striking distance of ground wires is larger than those of conductors, because the heights of ground wires are larger than those of conductors. Ground wires, therefore, attract more lightning than that predicted by the conventional electrogeometric models. In other words, shielding failures are less likely to occur for higher transmission lines, which is not the case as stated before.

One reason for the discrepancy in the shielding performance of UHV-designed transmission lines may be environmental conditions surrounding the lines. In the A-W model, the striking distance is determined by the lightning stroke current only—no environmental conditions that might affect the discharge paths are taken into account. Actual transmission lines are often subjected to fog or rain when lightning occurs nearby. Such conditions may affect the development of lightning discharges and cause the difference between the calculated lightning performance and field experiences.

The purpose of this study is to clarify the effect of the environment on the lightning shielding performance of UHV-designed transmission lines in foggy conditions, so long-gap discharge tests were conducted in dry and foggy conditions. Two types of test were conducted. First, rod-plane gap tests were conducted to analyze the fundamental specificity of discharge paths and the distortion of discharge paths was compared in dry and foggy conditions. Then, rod-conductor tests were conducted to analyze the features of lightning shielding the actual transmission lines using 1/50 scale models of typical UHV-designed transmission lines and

500kV transmission lines to compare the discharge rate of each conductor in dry and foggy conditions.

2. Test overview

To analyze the effect of fog, tests were conducted in dry and foggy conditions. Two types of tests were conducted. One was the rod-plane gap tests to evaluate the fundamental effect of fog on discharge, and the comparison of the eccentric distance of discharge paths was made in foggy and dry conditions. In the other tests, the conductors of UHV-designed transmission lines (a ground wire and three phases of power lines) were reduced in scale and simulated, and the frequency of discharge of each conductor was compared.

A. Test Conditions

1) Test site:

UHV fog chamber (dimensions: 35m×26m×35m) at the Yokosuka Research Laboratory, Central Research Institute of Electric Power Industry.

2) Applied voltage:

a) Waveform: switching impulse voltage waveform, 210/2400μs.

Previous reports describe the pulse-off time of a stepped leader as 37~124μs with a step length of 10~200m and a pulse-on time of 29~52μs with a step length of 3~50m. Therefore, the duration of one step of a stepped leader is 66~176μs, so the switching impulse was used with a duration closer to that of one step of a stepped leader than to that of the lightning impulse.

Meanwhile, to derive the equation for the striking distance proposed by Armstrong and Whitehead [$R_s = 6.72 \times I^{0.8}$ (m); I : lightning peak current magnitude (kA)], the discharge characteristics of the switching impulse in a long air gap were used in related references.

b) Polarity: negative, positive.

The applied waveforms were set to negative polarity to match most lightning strikes in Japan, and tests with a positive polarity were also conducted for reference purposes.

c) Applied voltage: 1.0 and 1.2 times the 50% flashover voltage.

With testing efficiency taken into account, 1.2 times the 50% flashover voltage was used in the tests using rod-plane electrodes, and 1.0 times the 50% flashover voltage was used in the rod-conductor tests.

B. Simulated Fog

To reproduce fog in the actual environment, cold fog was generated using water at room temperature. Fog spray nozzles at both ends of the discharge gap used high-pressure air and water to fill the discharge gap area between the high-voltage electrode and the ground electrode with fog. In order to completely fill the discharge gap with fog, a total of 16 nozzles were used, eight at each end. The water content of fog around the discharge gap during the test was 1.8 (g/m³).

3. Rod-plane electrode tests

A. Test Method

In order to analyze the specificity of discharge paths with and without fog, long-gap discharge tests were conducted by applying switching impulse voltages to rod-plane electrodes. Fig.3.5 shows the configuration of the rod-plane electrode tests. The gap length between the high-voltage rod electrode and the ground-plane electrode was 1.0m. The rod electrode was 10mm in diameter and 2m long. The plane electrode was 6m×6m (1cm mesh stainless steel). Discharge paths were photographed simultaneously by two cameras (cameras A and B) and discharge points on the plane electrode were determined in 3-D. The dispersion of the horizontal distances from the determined discharge points on the plane electrode to the rod electrode was evaluated over 25 to 30 repetitions of the test.

Fig. 3.6 shows the measurement method for discharge paths of rod-plane electrodes. The horizontal distance from the centerline, or the vertical straight line connecting the rod electrode and the plane electrode, were measured and evaluated. The eccentric distance at two locations was evaluated, one at the maximum eccentric position of the discharge path and the other at the discharge point.

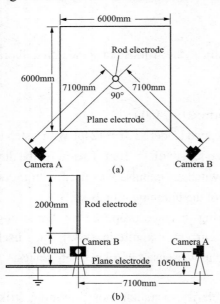

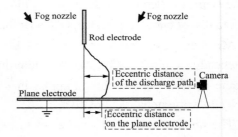

Fig.3.5　Configuration of the rod-plane electrode tests
(a) Plan view; (b) Elevation

Fig.3.6　Measurement of discharge paths of the rod-plane electrodes

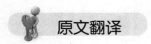

第二节　大雾条件下输电线路防雷中的气隙放电特性

1. 前言

在设计大规模输电线路时，雷电造成的停电是一个非常棘手的问题。常规的输电线路防雷设计所用到的是电气几何模型。杨等人最先利用下行先导通道和反击电流强度之间的联系建立了早期的电气几何模型。1968 年，阿姆斯壮和怀特黑德考虑到入射角度问题，对该电气模型进行了改进，并提出了一个相对简单的计算距离 R_s 的公式，如下所示

$$R_s = a \times I^b \tag{3.1}$$

式中：I 表示雷电流峰值；a 和 b 为常数。

常数 a 和 b 并不能通过理论得到，其值只能通过对比该模型计算出的输电线路故障率与现场实测的值来确定。实际上，虽然有其他文献提出关于常数 a 和 b 的不同值，但阿姆斯壮和怀特黑德提出的模型（A-W 模型）被广泛用于输电线路的防雷设计中。

然而，在日本的 500kV 超高压输电线路运行中，雷击故障发生次数却要多于采用 A-W 模型计算的故障次数。其原因在于所观察得到的雷击导线分布规律与 A-W 模型在设计阶段所假设的情况有很大的区别。

最近，越来越多的文献提出了更复杂的保护模型。这些文献所提出雷击距离不仅是反击电流的函数，同时也是接地物体高度的函数。根据这些模型，由于地线的高度一般大于其他导线，避雷地线的雷击距离大于其他导线。因此，地线吸引的雷击次数要比传统模型更多。此外，与前面所介绍的情况不同的是，在这种电压等级较高的输电线路中不太可能发生绕击事故。

造成超高压输电线路雷击保护效果不同的原因之一可能是线路周围的环境条件不同。在 A-W 模型中，雷击距离仅由雷击电流决定，缺乏对环境条件的考虑可能影响放电路径的确定。实际上雷电发生时输电线路经常遭受雨雾的影响。这种情况可能影响雷电放电的过程，导致线路防雷性能的计算值和实际值不同。

本文的研究目的是阐明在有雾的情况下，环境因素对超高压输电线路防雷特性的影响，故分别在干燥和有雾的情况下进行长气隙放电试验。分别按以下两种条件进行试验。首先，棒—板间隙试验旨在分析放电路径的基本原理，以及对比干燥和大雾条件下放电通路的失真情况。接下来，棒—极实验利用典型 500kV 超高压输电线路 1/50 比例模型来比较干燥和大雾条件下各导线的放电速度，旨在分析实际输电线路防雷特性。

2．试验概述

为了分析大雾的影响，分别在干燥和有雾的条件下进行试验。本文进行了以下两种类型试验。一种是棒—板间隙试验以评估雾对放电的基本影响，比较有雾和无雾条件下放电路径的偏心距。另一种是按特高压输电线路缩比模型（一根地线和三相导线）进行仿真，以比较各导线的放电频率。

A 试验条件

1）试验场地：

横须贺研究实验室在中央电力研究院超高压云雾室（尺寸规格：35m×26m×35m）。

2）外加电压：

a）波形：开关脉冲电压波形，210/2400μs。

梯级先导脉冲关断时间为 37～124μs，步长为 10～200m，脉冲启动时间为 29～52μs，步长为 3～50m。因此梯级先导单步持续时间为 66～176μs，故开关脉冲的持续时间更接近梯级先导的单步时间而不是雷电冲击时间。

同时，为了导出阿姆斯壮和怀特黑德提出的雷击距离的方程 [$R_s = 6.72 \times I^{0.8}$ (m)；I：雷电峰值电流（kA）]，需要利用长气隙开关的脉冲放电特性。

b）极性：负极性、正极性。

为符合日本大多数雷击情况，应用波形设置为负极性，同时也进行正极性试验以供参考。

c）外加电压：1.0 和 1.2 倍的 50%击穿电压。

考虑到测试效果，棒—板电极试验时采用 1.2 倍的 50%击穿电压，棒—极试验时采用 1.0 倍的 50%击穿电压。

B. 模拟大雾天气

为了重现真实条件下的大雾天气，采用常温下的水制成冷雾。在放电间隙终端的喷雾头采用高压空气和水来填补高电压电极和带雾地电极之间的放电气隙区域。为了让雾完全填满放电间隙，共采用16个喷头，一端8个。试验中放电气隙雾气含水量为1.8（g/m³）。

3. 棒—板电极试验

A. 试验方法

为了分别在有雾与无雾的条件下分析放电路径特性，对棒—板电极外加开关脉冲电压进行长气隙放电试验。图 3.5 显示了棒—板电极试验的结构。高压棒电极和接地板电极之间的间隙长度为 1m。棒电极直径为 10mm，长 2m。板电极尺寸为 6m×6m（1cm 不锈钢网）。放电路径被两个摄像头同时拍摄（相机 A 和 B），板电极上的放电点在三维空间测定。板电极的放电点到棒电极的放电点之间距离的散度要经过 25~30 次重复实验来进行评估。

图 3.6 示出了棒—电极放电路径的测量方法。该方法测量并评估了中心线或连接棒—电极垂线的距离。同时，该方法也评估了放电路径中最大偏心位置和其他放电点之间的偏心距离。

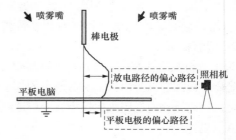

图 3.5 棒—板电极测试的结构
（a）俯视图；（b）主视图

图 3.6 棒—板电极放电路径的测量

New Words and Expressions

outage	*n.* 电力中断
distortion	*n.* 变形；失真
waveform	*n.* 波形
pulse	*n.* 脉冲
duration	*n.* 持续，持续的时间
flashover	*n.* 闪络
current magnitude	电流强度
ground wire	地线；避雷线

shielding failure	绕击
discharge path	放电路径
long-gap	长间隙
fog chamber	雾室
applied voltage	外加电压
impulse voltage	冲击电压；脉冲电压
stepped leader	梯级先导

 Notes

1. In other words, shielding failures are less likely to occur for higher transmission lines, which is not the case as stated before.

此外，与前面所介绍的情况不同的是，在这种高电压等级的输电线路中不太可能发生绕击事故。

2. Then, rod-conductor tests were conducted to analyze the features of lightning shielding the actual transmission lines using 1/50 scale models of typical UHV-designed transmission lines and 500kV transmission lines to compare the discharge rate of each conductor in dry and foggy conditions.

接下来，棒—极测试旨在分析实际输电线路防雷特性，利用典型 500kV 超高压输电线路 1/50 的缩比模型来比较干燥和大雾条件下各导线的放电速度。

3. In the other tests, the conductors of UHV-designed transmission lines (a ground wire and three phases of power lines) were reduced in scale and simulated, and the frequency of discharge of each conductor was compared.

另一种是按特高压输电线路缩比模型（一根地线和三相导线）进行仿真，以比较各导线的放电频率。

4. So the switching impulse was used with a duration closer to that of one step of a stepped leader than to that of the lightning impulse.

故开关脉冲的持续时间更接近梯级先导的单步时间，而不是雷电冲击时间。

 Exercises

Ⅰ. Choose the best answer into the blank.

1. In designing large-scale transmission lines, outages caused by ____ are a significant problem.

　　A. lightning　　　　B. wind　　　　C. fog　　　　D. rain

2. Actual transmission lines are often subjected to ____ or ____ when lightning occurs nearby.

　　A. fog; rain　　　　B. fog; wind　　　C. wind; rain　　D. rain; bird

3. To analyze the effect of fog, tests were conducted in ____ and ____ conditions.

A. snowy; rainy   B. dry; rainy   C. rainy; foggy   D. dry; foggy

4. In the other tests, the conductors of UHV-designed transmission lines were reduced in scale and simulated, and the _____ of discharge of each conductor was compared.

A. rate   B. path   C. frequency   D. efficiency

5. The duration of one step of a stepped leader is _____ μs.

A. 32～67   B. 66～176   C. 55～182   D. 132～201

II. Answer the following questions according to the text.

1. According to the models in this paper, why is the striking distance of ground wires larger than those of conductors?

2. In the A-W model, what are the striking distance not taken into account?

3. How many types of test were conducted in this paper?

4. Why did the tests use the 1.0 and 1.2 times the 50% flashover voltage?

5. What way did this paper think to analyze the specificity of discharge paths with and without fog?

III. Translate the following into Chinese.

Fig.2 shows the measurement method for discharge paths of rod-plane electrodes. The horizontal distance from the centerline, or the vertical straight line connecting the rod electrode and the plane electrode, were measured and evaluated. The eccentric distance at two locations was evaluated, one at the maximum eccentric position of the discharge path and the other at the discharge point.

Keys

I. Choose the best answer into the blank.

1. A   2. A   3. D   4. C   5. B

II. Answer the following questions according to the text.

1. Because the heights of ground wires are larger than those of conductors.

2. In the A-W model, the striking distance is determined by the lightning stroke current only—no environmental conditions that might affect the discharge paths are taken into account.

3. Two.

4. Because the tests taking testing efficiency into account, 1.2 times the 50% flashover voltage was used in the tests using rod-plane electrodes, and 1.0 times the 50% flashover voltage was used in the rod-conductor tests.

5. In order to analyze the specificity of discharge paths with and without fog, long-gap discharge tests were conducted by applying switching impulse voltages to rod-plane electrodes.

III. Translate the following into Chinese.

图 2 显示了棒—板电极放电路径的测量方法。测量和评估到中心线的水平距离或者连接棒电极和板电极的垂线距离。在两个地点进行偏心距评估，一是在放电通路中的最大偏心位

置，另一个是在放电点。

翻译技巧之翻译方法（一）

1. 直译与意译

直译指基本保留原有句子结构，照字面意思翻译；意译是在不损害原文内容和精神的前提下，为了表达的需要，对原文做相应的调整。如：

Good marriages don't just happen. They take a lot of love and a lot of work.

直译：好的婚姻不会仅仅发生——它们需要大量的爱和大量的工作。

意译：幸福的婚姻不是凭空发生的——它需要你为它付出大量的爱和做大量的工作。

或：美满的婚姻不会从天上掉下来——你必须为它付出大量的爱，做大量的工作。

很显然，本句话的意译要比直译更符合汉语表达习惯。当然，一句话并不限于一种译法，要根据具体需要而定。一般来说，在英汉翻译考试中，如果直译能达意就用直译，如果直译效果不好，就应该考虑意译。只要译文内容忠实，意思明白就行了。

2. 顺译法（又名句型对应法）

顾名思义，顺译法（句型对应法）就是按原文句子结构的排列顺序进行翻译，这种译法适合于原文叙述层次与汉语相近的长句翻译，如只含名词性从句的复合句、前置的状语从句或从句在后的长复合句等。

As an obedient son, I had to accept my parents' decision that I was to be a doctor, though the prospect interested in me not at all. 作为一个孝顺的儿子，我不得不接受父母的决定，去当大夫，虽然我对这样的前途毫无兴趣。

3. 倒译法

倒译法就是颠倒原文句子结构的排列顺序来进行翻译。

The moon is completely empty of water because the gravity on the moon is much less than on the earth. 因为月球的引力比地球小得多，所以月球上根本没有水。

The football students can be removed from the university if they fail to pass their examination. 作为足球运动员的学生如果考试不及格就要被开除。

Section 3 Short-circuit Current

1. Terms and Definitions

The following terms and definitions correspond largely to those defined in IEC 60909. Refer to this standard for all terms not used in this paper.

The terms short circuit and ground fault describe faults in the isolation of operational equipment which occur when live parts are shunted out as a result.

(1) Causes:

1) Overtemperatures due to excessively high overcurrents.

2) Disruptive discharges due to overvoltages.

3) Arcing due to moisture together with impure air, especially on insulators.

(2) Effects:

1) Interruption of power supply.

2) Destruction of system components.

3) Development of unacceptable mechanical and thermal stresses in electrical operational equipment.

(3) Short circuit:

According to IEC 60909, a short circuit is the accidental or intentional conductive connection through a relatively low resistance or impedance between two or more points of a circuit which are normally at different potentials.

(4) Short circuit current:

According to IEC 60909, a short circuit current results from a short circuit in an electrical network.

It is necessary to differentiate here between the short circuit current at the position of the short circuit and the transferred short circuit currents in the network branches.

2. Short circuit path in the positive-sequence system

For the same external conductor voltages, a three-pole short circuit allows three currents of the same magnitude to develop between the three conductors. It is there for only necessary to consider one conductor in further calculations. Depending on the distance from the position of the short circuit from the generator, here it is necessary to consider near-to-generator and far-from-generator short circuits separately.

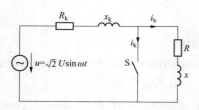

Fig.3.7 Equivalent circuit of the short circuit current path in the positive-sequence system

For far-from-generator and near-to-generator short circuits, the short circuit path can be represented by a mesh diagram with AC voltage source, reactances X and resistances R (Figure.3.7). Here, X and R replace all components such as cables, conductors, transformers, generators and motors.

The following differential equation can be used to describe the short circuit process

$$i_k \times R_k + L_k \frac{di_i}{dt} = \hat{u} \times \sin(\omega t + \psi) \quad (3.2)$$

Where ψ is the phase angle at the point in time of the short circuit; ω is the angular frequence.

This assumes that the current before S closes (short circuit) is zero. The inhomogeneous first order differential equation can be solved by determining the homogeneous solution i_k and a particular solution i_{k-}^2.

$$i_k = i_{k-}''' + i_{k-} \quad (3.3)$$

Where i_{k-} is the general solution.

Chapter 3 Corona Discharge

The homogeneous solution, with the time constant g = L/R, solution yields

$$i_k' = \frac{-\hat{u}}{\sqrt{(R^2+X^2)}} e^{\frac{t}{\tau_g} \sin(\psi-\varphi_k)} \tag{3.4}$$

For the particular solution, we obtain

$$i_k'' = \frac{-\hat{u}}{\sqrt{(R^2+X^2)}} \sin(\omega t + \psi - \varphi_k) \tag{3.5}$$

The total short circuit current is composed of both components

$$i_k = \frac{-\hat{u}}{\sqrt{(R^2+X^2)}} \left[\sin(\omega t + \psi - \varphi_k) - e^{\frac{t}{\tau_g}} \sin(\psi - \varphi_k) \right] \tag{3.6}$$

The phase angle of the short circuit current (short circuit angle) is then, in accordance with the above equation

$$\varphi_k = \psi - v = \arctan\frac{X}{R} \tag{3.7}$$

For the far-from-generator short circuit, the short circuit current is therefore made up of a constant AC periodic component and the decaying DC aperiodic component. From the simplified calculations, we can now reach the following conclusions:

(1) The short circuit current always has a decaying DC aperiodic component in addition to the stationary AC periodic component.

(2) The magnitude of the short circuit current depends on the operating angle of the current. It reaches a maximum at c = 90°(purely inductive load). This case serves as the basis for further calculations.

(3) The short circuit current is always inductive.

3. Methods of short circuit calculation

There are three methods of calculation for short circuit current in three phase system:

(1) Calculating the equivalent voltage source in fault location.

(2) Determining the load flow situation by superposition method.

(3) Transient calculation.

4. Calculating with reference variables

There are several methods for performing short circuit calculations with absolute and reference impedance values. A few are summarized here and examples are calculated for comparison. To define the relative values, there are two possible reference variables.

For the characterization of electrotechnical relationships we require the four parameters:

(1) Voltage U in V;

(2) Current I in A;

(3) Impedance Z in Ω;

(4) Apparent power S in VA.

Three methods can be used to calculate the short circuit current:

(1) The Ohm system: Units is kV, kA, V, MVA.

(2) The PU system: This method is used predominantly for electrical machines; all four parameters u, i, z and s are given as per unit ($unit = 1$). The reference value is 100MVA. The two reference variables for this system are UB and SB. Example: The reactances of a synchronous machine X_d, X_d', X_k'' are given in PU or in % PU, multiplied by 100 %.

(3) The %/MVA system: This system is especially well suited for the fast determination of short circuit impedances. As formal unit only the % symbol is add.

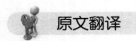原文翻译

第三节 短 路 电 流

1. 术语和定义

下列术语和定义基本上与 IEC 60909 中定义的术语和定义相一致。未出现在本文中的术语可以在该标准中查询。

短路和接地故障主要是操作设备过程中带电部分被切换而导致绝缘损坏的结果。

（1）原因：

1）温度过高导致强烈的过电流。

2）火花放电导致过电压。

3）电弧是由潮湿、带有杂质的空气引起的，常发生于绝缘子上。

（2）后果：

1）供电中断。

2）系统组成部分被破坏。

3）电气设备所承受的有害机械力和热应力增大。

（3）短路：

根据 IEC 60909，短路是通过相对低的电阻连接，或两个不同电位之间的意外或故意的导电连接而导致的。

（4）短路电流：

根据 IEC 60909，短路电流是电力网络短路而产生的。在这里有必要区分不同短路位置的短路电流和网络分支中的转移短路电流。

2. 正序分量的短路路径

对于导体具有相同外部电压的条件来说，三相导体中发生三相短路而产生的三相短路电流在每一相的大小是相同的。所以在下一步的计算中只需要考虑一相的情况。根据短路点到发电机距离的不同，将远离发电机短路和靠近发电机短路分为两种情况进行考虑。

对于远离或靠近发电机短路的情况，短路回路可以用一个有交流电压源、电抗 X、电阻 R 的网络图表示（图 3.7）。X 和 R 可替代所有的元件，如电缆、导体、变压器、发电机和电机。

微分方程可以用来描述短路过程

$$i_k \times R_k + L_k \frac{di_i}{dt} = \hat{u} \times \sin(\omega t + \psi) \quad (3.2)$$

式中：ψ是短路电流的相位角；ω是角频率。

假设电流在 S 关闭（短路）之前是零。一阶非齐次微分方程的解可通过求解方程的齐次解 i_k 和特解 i^2_{k-}。

$$i_k = i''_{k-} + i_{k-} \quad (3.3)$$

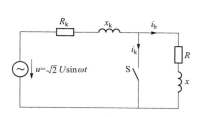

图 3.7 短路电流路径在正序系统中的等效电路

式中：i_{k-} 为通解。

齐次解的时间常数 $g = L/R$，方程式为

$$i_k = \frac{-\hat{u}}{\sqrt{(R^2 + X^2)}} e^{\frac{t}{\tau_g} \sin(\psi - \varphi_k)} \quad (3.4)$$

对于特解，可得

$$i''_k = \frac{-\hat{u}}{\sqrt{(R^2 + X^2)}} \sin(\omega t + \psi - \varphi_k) \quad (3.5)$$

总的短路电流由两部分构成

$$i_k = \frac{-\hat{u}}{\sqrt{(R^2 + X^2)}} \left[\sin(\omega t + \psi - \varphi_k) - e^{\frac{t}{\tau_g}} \sin(\psi - \varphi_k) \right] \quad (3.6)$$

根据以上方程，单相短路电流的相位角为

$$\varphi_k = \psi - \nu = \arctan \frac{X}{R} \quad (3.7)$$

对于远离发电机的短路，短路电流是由一个不变的交流周期分量和一个衰减的直流非周期分量构成。简化计算后，我们可以得出以下结论：

（1）短路电流总是由一个衰减的直流非周期分量和一个固定的交流周期分量构成。

（2）短路电流的大小取决于电流的相位，最大值在相位为 $c=90°$（纯电感负载）时取得。这种情况作为进一步计算的基础。

（3）短路电流总是感应得到的。

3. 短路电流计算方法

三相系统中的短路电流有三种计算方法：

（1）在故障位置计算等效电压源。

（2）通过迭代法确定潮流情况。

（3）暂态计算。

4. 参考变量的求解

有很多方法可根据绝对阻抗值和参考阻抗值进行短路计算。本文已经总结出几种计算方法，并引用相关算例加以对比。为了定义相对值，有两种可能的参考变量定义。

为表征电气关系，我们要求四个参数：

（1）单位为 V 的电压 U；

（2）单位为 A 的电流 I；

（3）单位为Ω的阻抗 Z；

（4）单位为 VA 的视在功率 S。

以下有三种方法可用来计算短路电流：

（1）欧姆系统：单位为 kV、kA、V、MVA。

（2）标幺值系统：这种方法主要用于发电机系统，所有四个参数，u、i、z 和 s 都被设定为 1。参考容量设为 100MVA。系统的两个基准变量分别是 UB 和 SB。例如：同步机的电抗 X_d、X_d'、X_k'' 都以标幺值的形式给定，或者乘以 100%以百分比标幺值形式给定。

（3）%/MVA 系统：这种方法适用于短路阻抗的快速计算。在标准单位中仅添加了%符号。

 New Words and Expressions

component	n. 元件
resistance	n. 电阻
impedance	n. 阻抗
reactance	n. 电抗
cable	n. 电缆
electrotechnical	adj. 电工的
ohm	n. 欧姆（电阻单位）
three-pole	n. 三极
equivalent voltage source	等效电压源
ground fault	［电］接地故障
overcurrent	n. 过负荷电流；过量电流
electrical	adj. 有关电的；电气科学的
conductive	adj. 传导的；传导性的；导电
motor	n. 发动机

 Notes

1. It is necessary to differentiate here between the short circuit current at the position of the short circuit and the transferred short circuit currents in the network branches.

在这里有必要区分在短路位置的短路电流和网络分支中的转移短路电流。

2. For far-from-generator and near-to-generator short circuits, the short circuit path can be represented by a mesh diagram with AC voltage source, reactances X and resistances R.

对于远离或靠近发电机短路的情况，短路路径可以用一个有交流电压源、电抗 X、电阻 R 构成的网络图表示。

3. For the far-from-generator short circuit, the short circuit current is therefore made up of a constant AC periodic component and the decaying DC aperiodic component.

对于远离发电机形式的短路，短路电流由一个不变的交流周期分量和一个衰减的直流非周期分量构成。

4. To define the relative values, there are two possible reference variables.

为了定义相对值，有两种可能的参考变量。

Exercises

Ⅰ. Choose the best answer into the blank.

1. Which reason couldn't led to the short circuit and ground fault? ____
   A. overtemperatures.　　　　　　B. disruptive discharges.
   C. arcing.　　　　　　　　　　　D. interruption of power supply.
2. How many effects if short circuit and ground fault happened? ____
   A. 4　　　　　B. 3　　　　　C. 2　　　　　D. 1
3. In methods of short circuit calculation, which is not include? ____
   A. Calculating the equivalent voltage source in fault location.
   B. Calculating the equivalent circuit source in fault location.
   C. Determining the load flow situation by superposition method.
   D. Transient calculation.
4. For the characterization of electrotechnical relationships we require the four parameters which of the following is correct? ____
   A. Voltage U in kV.　　　　　　B. Current I in mA.
   C. Impedance Z in Ω.　　　　　D. Apparent power S in VA.

Ⅱ. Answer the following questions according to the next.

1. According to IEC 60909, What is the definition of a short circuit current?
2. From the simplified calculations, what conclusions can we reach the following for the far-from-generator short circuit.
3. Please list three methods for performing short circuit calculations with absolute and reference impedance values.

Ⅲ. Translate the following into Chinese.

The PU system: This method is used predominantly for electrical machines; all four parameters U, I, Z and S are given as per unit (*unit* = 1). The reference value is 100MVA. The two reference variables for this system are UB and SB. Example: The reactances of a synchronous machine X_d, X_d', X_k'' are given in PU or in % PU, multiplied by 100 %.

Keys

Ⅰ. Choose the best answer into the blank.

1. D　　2. B　　3. B　　4. D

Ⅱ. Answer the following questions according to the next.

1. According to IEC 60909, a short circuit current results from a short circuit in an electrical network.

2. 1) The short circuit current always has a decaying DC aperiodic component in addition to the stationary AC periodic component.

2) The magnitude of the short circuit current depends on the operating angle of the current. It reaches a maximum at $c = 90°$ (purely inductive load). This case serves as the basis for further calculations.

3) The short circuit current is always inductive.

3. 1) The Ohm system; 2) The PU system; 3) The %/MVA system.

Ⅲ. Translate the following into Chinese.

标幺值系统：这种方法主要用于发电机系统，所有四个参数，U、I、Z 和 S 都被设定为 1。参考容量设为 100MV。系统的两个参考变量分别是 UB 和 SB。例如：同步机的电抗 X_d、X_d'、X_k'' 都以标幺值的形式被给定，或者乘以 100% 以百分比标幺值形式给定。

翻译技巧之翻译方法（二）

1. 分译法

分译法，又称拆译法，也是一种基本的句法变通手段。从被分译成分的结构而言，分译大致可以分为单词的分译、短语的分译和句子的分译三种。

（1）单词的分译即拆词，将难译的词从句子主干中拆离出来，另做处理，这种方法常常引起句式上的调整，英译汉中要拆译的词常常是形容词和副词。如：

He unnecessarily spent a lot of time introducing his book, which the student are familiar with. 他花了很长时间介绍这本书，其实没有必要，因为学生们对它已经很熟悉了。

Strange enough they were the same age to the day. 说来也巧，他俩年纪一样大，而且还是同日的。

（2）短语的分译是指把原文中的一个短语分译成一个句子。名词短语、分词短语、介词短语等有时都可以分译成句。如：

These cheerful little trams, dating back to 1873, chug and sway up to towering hills with bells ringing and people hanging from every opening. 这些令人欢快的小缆车建于 1873 年，咔嚓咔嚓地摇摆爬上高耸的山峦。车上铃儿叮当作响，每个窗口都是人。（介词短语分译）

Invitingly green Angel Island, once a military installation, contains meandering trails and picnic spots ideal for a day's excursion. 迷人的天使岛郁郁葱葱，小径蜿蜒，是一日游的理想野餐场所。但在过去它却是一个军事基地。（名词短语分译）

（3）句子的分译分为两种：结构分译和语义分译。结构分译是指原文有些句子的结构十分奇特，其主、谓、宾汉译时无法对号入座。这时应从整个句子着手，打散整个结构，重新翻译。如：

What can easily be seen in his poems are his imagery and originality, power and range. 他的诗形象生动，独具一格，而且气势磅礴，题材广泛。这是显而易见的。

原句中的某一部分，汉译时从句义来看，与其放在原句，倒不如放在下一句中更合适。

这时，可把有关部分放到译文的另一个句子里。

She used to relate how she met in Italy an elderly gentleman who was looking very sad. She inquired the cause of his melancholy and he said that he said that he had just parted from his two gradchildern. 她曾提到她在意大利碰到一位老先生的事。她见他愁容满面，就问起他闷闷不乐的原因。他说那是因为他刚同两个孙子分手。

2. 合译法

合译是将原文的两个或几个分开叙述的意思或层次合并重组，如将两个分句合译为一个简单句，或两个简单句合译成一个复合句等，使全句的结构更加紧凑，语气更加顺通。

There are men here from all over the country. Many of them are from the north. 从全国各地来的人中有许多是北方人。

There are many people who want to see the film. 许多人要看这部电影。

The time was 10:30, and traffic on the street was light. 10 点 30 分的时候，街上来往的车辆稀少了。

Chapter 4　The Design of Power Transmission Line

Section 1　Overview of the Transmission Line Design Process

An overview is presented for the design of overhead high-voltage transmission lines. Since most lines constructed by Los Angeles Department of Water and Power in recent years have been extra-high-voltage AC, this paper emphasizes the design of 500kV AC transmission lines built in the southwest United States. Empirically developed practices are presented throughout the paper.

1. Introduction

Los Angeles Department of Water and Power (LADWP) has transmission lines dating from 1915 until the present, and new lines are almost always being planned. For the AC transmission lines, the voltages range from 115 to 500kV with the number of circuits ranging from one to six per tower.

The design process used by LADWP in constructing high-voltage AC transmission lines is presented in this paper. The design process begins with the required engineering data, then continues with sections on conductor selection, wind and ice loading, maximum tension determination, transmission towers, insulators, ruling span determination, spotting towers, magnetic field effects and environmental criteria.

2. Engineering Data

The engineering data are obtained from the Transmission Design, Transmission Planning, Resource Planning and Environmental and Governmental Affairs sections of LADWP. Typical design data are outlined below:

(1) transmission voltage;

(2) levelized current value;

(3) amortized life (typically 40~50 years);

(4) location of transmission line corridor;

(5) environmental constraints.

3. Conductors

Historically, LADWP has used standard round-rod steel reinforced aluminum conductor (ACSR). The reason for ACSR's popularity is its low relative cost and its high weight-to-strength ratio as compared with other conductor material. In addition, ACSR is commercially available in wide ranges of mechanical strength and electric current capacity ratings.

The most economical conductor size is determined using Kelvin's law, which states: "The most economical area of a conductor is that for which the annual cost of the energy losses is equal to the interest on that portion of the capital outlay which may be considered as proportional to the weight of the conductor".

Typically, conductor selection for a new high-voltage transmission line would be as follows.

(1) A number of different candidate conductors (or candidate conductor bundles) are selected based on minimum operational requirements. Minimum operational requirements include thermal capacity (i.e. capacity to carry the maximum anticipated line current for specified time periods and ambient conditions) and maximum allowable voltage gradient effects (i.e. corona, radio interference, television interference and audible noise limitations). Conductor temperature-sag-tension characteristics may also be considered.

(2) For each candidate conductor (or conductor bundle) estimates are made of the line construction (installation) cost and the present worth of the I^2R losses over the projected life of the line. These costs are added to obtain the total estimated cost associated with the installation and use of each candidate conductor. The candidate conductor with the lowest total estimated cost is the selected conductor.

4. Wind and ice loading

A wind and ice loading estimate is necessary to determine at what conditions the loaded design tension will occur. For California, General Order 95 (G.O. 95), Rule 43 on temperature and loading applies. For the rest of the US the National Electrical Safety Code (NESC) Section 23 on clearances applies, as well as any regulations imposed by the state.

5. Maximum tension determination

Based on the conductor type selected, the maximum (loaded) design tension is determined. The loaded design tension (i.e. the tension of the conductor with wind and ice loading) is typically specified to be 33%~35% of the ultimate strength under loaded conditions. Unloaded conductor sagged-in tension [60°F (15.5°C), no wind, no ice] is typically 20%~22% of the ultimate strength.

For example, assume the conductor selected was a 2312 kcmiP ACSR "Thrasher". The ultimate strength of a "Thrasher" conductor is approximately 55 000~60 000lbf (245~267kN) depending on the galvanization of the core. The loaded design tension is 33% of 60 000lbf, i.e. 20 000lbf (89kN).

6. Transmission towers

Tower loading information based on the loaded conductor and overhead ground wire design tensions, along with the conductor-to-steel electrical clearance, is given to the structural design engineer to design the towers. Historically, LADWP towers used for AC lines are of the lattice steel self-supporting type. Other utilities sometimes use guyed towers which have a lower weight than the self-supporting type, and this weight can be further reduced by the use of aluminum instead of steel.

7. Insulators

The basic electrical requirement of any insulator assembly is that the insulator should sustain a lightning strike without forming a conductor path over the insulator surface. The basic mechanical requirement is that the insulator be able to support the wind and ice loading of the conductor and to withstand lightning, power surges and mechanical abuse without dropping the conductor.

The operational performance of a transmission line is largely determined by the insulation. Selecting the insulation levels requires careful analysis. Insulators near refineries, along the seashore or in areas of light rainfall may become so contaminated that considerable overinsulation is required. Under normal conditions, the assembled insulator should have a dry flashover of five times the nominal operating voltage and a leakage path of twice the shortest air-gap distance.

The standard insulator is a porcelain insulator with a 10 in. (25.4cm) diameter and a spacing of $5\frac{3}{4}$ in. (14.6cm) from center to center. In recent years silicone polymer insulators have been installed in many new lines. The advantage of using polymer insulators is that contamination is less of a problem than with porcelain insulators, which reduces line maintenance. The disadvantage of polymer insulators is that they deteriorate much faster than porcelain and need replacement sooner.

原文翻译

第一节　输电线路设计过程简介

本文将对高压架空输电线路设计进行简单的介绍。由于最近几年洛杉矶水利电力部建造的输电线路大部分是超高压交流输电线路，因此，本文将对美国西南部500kV交流输电线路的设计情况进行着重介绍。并且在介绍过程中，会引用较多成熟的工程案例。

1. 引言

洛杉矶水利电力部（LADWP）从1915年开始建造输电线工程，且现在仍然在不断地规划新的输电线路。对于交流输电线路来说，115～500kV电压等级的输电线路在每一基铁塔中有1～6回输电线路。

本文将介绍洛杉矶水利电力部在建设高压交流输电线路时的设计过程。设计时，首先要得到所需的工程数据，接着就可以选择导线，计算风荷载、覆冰荷载和最大极限张力，确定铁塔、绝缘子、线路挡距和定位塔，得到输电线路的电磁环境及其限值。

2. 工程数据

工程数据可以从洛杉矶水利电力部的输电线路设计、输电线路规划、资源规划及环境与行政部门中获得。标准的设计数据主要有以下几个方面：

（1）线路电压等级；

（2）平均电流值；

（3）分期还款年限（一般是40～50年）；

（4）输电线路走廊的位置；

（5）环境限值。

3. 导线

到目前为止，洛杉矶水利电力部一直都在使用标准钢芯铝绞线（ACSR）。钢芯铝绞线一直被使用的原因在于，钢芯铝绞线重量与强度的比值比其他材料的导线更高，而成本却相对较低。此外，钢芯铝绞线是在市场上可以买到的机械强度和额定载流量规格较多的产品。

确定导线尺寸最经济的方法是采用开尔文定律，即"使每年由电能损失造成的费用和与导线重量成正比的基建贷款的利息相等。"

一般来说，新建高压输电线路的导线选择方法如下：

（1）在许多不同的备选导线（包含分裂导线）中，根据最低运行要求进行选择。最低运行要求包括热稳定性（即在特定时间和环境下能够达到最大预期线路电流的能力）和最大允许电压梯度（即电晕、无线电干扰、电视干扰和可听噪声造成的限制），以及导体温度、弧垂和张力特性。

（2）每个备选导线（包含分裂导线）的预估成本是由架设（安装）线路的成本和使用寿命内的电能损耗费用构成。与每个备选导线安装和使用有关的这些预估成本相加可获得总预估成本。选择总预估成本最低的备选导线。

4. 风荷载和覆冰荷载

预估风荷载和覆冰荷载是非常必要的，因为这些荷载决定了最大设计张力在何种气象条件下发生。加州采用了 95 号法规中第 43 条关于温度和荷载的规定。美国其他州则采用了国家电气安全规范第 23 节，这个规范和美国其他法律具有同等地位。

5. 最大张力的确定

根据所选导线的类型可以确定最大设计张力。有荷载情况下的最大设计张力（即在风荷载和覆冰荷载情况下导线的张力）通常是该荷载条件下导线拉断力的 33%～35%。无荷载情况下 [60°F（15.5℃）、无风、无冰] 导线的最大设计张力通常是导线拉断力的 20%～22%。

例如，假设所选导线是一个截面积为 2312 千圆密耳的钢芯铝绞线"思拉舍"。一根"思拉舍"导线的拉断力由芯层决定，为 55 000～60 000 磅（245～267kN）。有荷载情况下的设计张力为 60 000 磅的 33%，即 20 000 磅（89kN）。

6. 输电杆塔

有荷载情况下导线和架空地线的设计张力，以及导线到角钢的电气间隙所确定的杆塔负载信息要提供给结构设计工程师，用来设计杆塔。到目前为止，洛杉矶水利电力部用于交流输电线路的杆塔一般为桁架形的钢结构自立式杆塔。其他公用设施有时使用重量低于自立式杆塔的拉线塔，若将钢由铝材代替可进一步降低重量。

7. 绝缘子

所有绝缘子串的基本电气要求是在遭受雷击情况下，绝缘子串表面不形成导电回路。基本的机械要求是能够承受导线的风荷载和覆冰荷载，并能承受雷电、电涌和机械损伤而不会松脱导线。

输电线路的运行性能很大程度上取决于绝缘。因此选择绝缘水平要仔细分析。炼油厂附近、沿海或降雨较少的地区，其绝缘子可能被严重污染，因此这些地区的绝缘要求较高。正常情况下，绝缘子串应在五倍标称工作电压和两倍最短空气间隙距离的泄漏路径下发生干闪。

标准的绝缘子是一个直径为 10 英尺（25.4cm），两片绝缘子间距为 5 英尺（14.6cm）的瓷绝缘子。近年来，许多新的输电线路已安装有机硅复合绝缘子。使用复合绝缘子的优点是污闪问题比瓷绝缘子少，降低了线路维护的工作量；复合绝缘子的缺点是老化速度瓷绝缘子快，需要频繁更换。

New Words and Expressions

circuit	n. 电路，回路
magnetic	adj. 地磁的；有磁性的
magnetic field	磁场
steel reinforced	钢骨
voltage gradient	电压梯度
radio interference	无线电干扰
mechanical strength	机械强度
capacity rating	额定功率；额定容量
power surge	电涌
porcelain insulator	瓷绝缘子
silicone polymer insulator	有机复合绝缘子
ruling span	代表挡距
spoting tower	杆塔定位

Notes

1. The design process begins with the required engineering data, then continues with sections on conductor selection, wind and ice loading, maximum tension determination, transmission towers, insulators, ruling span determination, spotting towers, magnetic field effects and environmental criteria.

本文将介绍洛杉矶水利电力部在建设高压交流输电线路时的设计过程。设计时，首先要得到所需的工程数据，接着就可以选择导线，计算风荷载、覆冰荷载和最大极限张力，确定铁塔、绝缘子、线路挡距和定位塔，得到输电线路的电磁环境及其限值。

2. The most economical conductor size is determined using Kelvin's law, which states: "The most economical area of a conductor is that for which the annual cost of the energy losses is equal to the interest on that portion of the capital outlay which may be considered as proportional to the weight of the conductor".

确定导线尺寸最经济的方法是采用开尔文定律，即"使每年由电能损失造成的费用和与导线重量成正比的基建贷款的利息相等。"

3. The basic electrical requirement of any insulator assembly is that the insulator should sustain a lightning strike without forming a conductor path over the insulator surface.

任何绝缘子串的基本电气要求是在遭受雷击情况下，绝缘子串表面不形成导电回路。

Exercises

I. Choose the best answer into the blank.

1. For the AC transmission lines, which voltage is not include ___?
   A. 115kV   B. 350kV   C. 450kV   D. 550kV
2. Which is NOT the typical design date above___ ?
   A. transmission voltage   B. magnetic field
   C. levelized current value   D. environment constraints
3. What is the basic electrical requirement of any insulator assembly ___ ?
   A. sustain a lightning strike   B. support the wind
   C. withstand mechanical abuse   D. withstand power surges
4. What is the operational performance of a transmission line largely determined by ___?
   A. the insulation   B. the capacity rating
   C. material of lines   D. the working environment

II. Answer the following questions according to the next.

1. Please describe the design process used by LADWP in constructing high-voltage AC transmission lines is presented in this paper in short.

2. How to use the Kelvin's law to determine the most economical conductor size?

3. Typically, how to select conductor for a new high-voltage transmission line?

4. What is the standard insulator?

5. Please talk about the advantage and disadvantage of using polymer insulations.

III. Translate the following into Chinese.

Based on the conductor type selected, the maximum (loaded) design tension is determined. The loaded design tension (i.e. the tension of the conductor with wind and ice loading) is typically specified to be 33%～35% of the ultimate strength under loaded conditions. Unloaded conductor sagged-in tension [60°F (15.5℃), no wind, no ice] is typically 20%～22% of the ultimate strength.

For example, assume the conductor selected was a 2312 kcmiP ACSR "Thrasher". The ultimate strength of a "Thrasher" conductor is approximately 55 000～60 000lbf (245～267kN) depending on the galvanization of the core. The loaded design tension is 33% of 60 000lbf, i.e. 20 000lbf (89 kN).

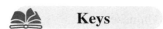

Keys

I. Choose the best answer into the blank.
1. D   2. B   3. A   4. A

II. Answer the following questions according to the next.

1. The design process begins with the required engineering data, then continues with sections on conductor selection, wind and ice loading, maximum tension determination, transmission towers, insulators, ruling span determination, spotting towers, magnetic field effects and environmental criteria.

2. The most economical area of a conductor is that for which the annual cost of the energy losses is equal to the interest on that portion of the capital outlay which may be considered as proportional to the weight of the conductor.

(1) A number of different candidate conductors (or candidate conductor bundles) are selected based on minimum operational requirements. Minimum operational requirements include thermal capacity (i.e. capacity to carry the maximum anticipated line current for specified time periods and ambient conditions) and maximum allowable voltage gradient effects (i.e. corona, radio interference, television interference and audible noise limitations). Conductor temperature-sag-tension characteristics may also be considered.

(2) For each candidate conductor (or conductor bundle) estimates are made of the line construction (installation) cost and the present worth of the I^2R losses over the projected life of the line. These costs are added to obtain the total estimated cost associated with the installation and use of each candidate conductor. The candidate conductor with the lowest total estimated cost is the selected conductor.

3. (1) A number of different candidate conductors (or candidate conductor bundles) are selected based on minimum operational requirements. Minimum operational requirements include thermal capacity (i.e. capacity to carry the maximum anticipated line current for specified time periods and ambient conditions) and maximum allowable voltage gradient effects (i.e. corona, radio interference, television interference and audible noise limitations). Conductor temperature-sag-tension characteristics may also be considered.

(2) For each candidate conductor (or conductor bundle) estimates are made of the line construction (installation) cost and the present worth of the I^2R losses over the projected life of the line. These costs are added to obtain the total estimated cost associated with the installation and use of each candidate conductor. The candidate conductor with the lowest total estimated cost is the selected conductor.

4. The standard insulator is a porcelain insulator with a 10 in. (25.4cm) diameter and a spacing of $5^{3/4}$ in. (14.6cm) from center to center.

5. The advantage of using polymer insulators is that contamination is less of a problem than with porcelain insulators, which reduces line maintenance. The disadvantage of polymer insulators is that they deteriorate much faster than porcelain and need replacement sooner.

III. Translate the following into Chinese.

根据所选导线的类型可以确定最大设计张力。有荷载情况下的最大设计张力（即在风荷载和覆冰载荷情况下导线的张力）通常是该载荷条件下导线拉断力的 33%～35%。无荷载情况下［60°F（15.5℃）、无风、无冰］导线的最大设计张力通常是导线拉断力的 20%～22%。

例如，假设所选导线是一个截面积为 2312 千圆密耳的钢芯铝绞线"思拉舍"。一根"思拉舍"导线的拉断力由芯层决定，为 55 000～60 000 磅（245～267kN）。有荷载情况下的设计张力为 60 000 磅的 33%，即 20 000 磅（89kN）。

翻译技巧之被动句（一）

在对科技英语进行翻译时要注意其特点，它不像翻译文学作品，要求译出原作者的风格、笔调，保持原作的艺术形象。翻译时应遵循"忠实"和"通顺"两个准则。"忠实"即译文应忠实于原文，准确、完整、科学地表达原文的内容，译者不得任意对原文内容加以歪曲、增删、遗漏和篡改，不做主观渲染，不用带感情色彩的词汇。"通顺"是说译文应当符合中文的语法要求，使读者看起来通俗易懂。在科技翻译中。要达到融会贯通，必须了解相关的科技知识，掌握同一事物的中英文表达方式。单纯靠对语言的把握也能传达双方的语言信息。但运用语言的灵活性特别是选词的准确性会受到很大限制。要解决这个问题，翻译人员就要积极主动地熟悉这个科技领域的相关知识。

我们在了解科技英语中被动语态的特点和翻译准则之后可以采取译成汉语主动句、译成汉语被动句和译成汉语无主句的翻译方法：

译成汉语主动句的介绍内容如下：

（1）用"人们""大家""有人""我们"等含有泛指意义的词作主语，从而使汉语译文呈"兼语式"句式。

The mechanism of fever production is not completely understood. 人们还不完全清楚发烧的产生机理。

Sun is known to rise in the east. 我们都知道太阳从东方升起。

（2）由 it 引导的主语从句，译成无人称或不定人称句。

It is said that the production of transistor radios Was increased six times from 1970 to 1974. 据说从 1970～1974 年，晶体管收音机的产量增长了五倍。

It is demonstrated that the conductivity of silver is higher than that of copper. 已经证明银的导电率比铜的导电率高。

这类句型还有：

It is reported/supposed/must be admitted/pointed out that　据报道/据推测/必须承认/必须指出

It is believed/generally considered/well-known that　有人相信/大家认为/众所周知

（3）当被动语态中的主语为无生命的名词或由 under、in、on 等少数几个介词短语构成时，译成主动句。

That computer is under repair. 那台计算机正在修理。

A new design method is on trial. 这种新的设计方法正在试用。

This kind of electronic equipment is in great demand. 这种电子设备需求量很大。

The plan is being carried out. 计划正在实施。

（4）当 need、want、require 等动词后接主动形式的动名词，表示被动意义时，译成主动句。

This device needs repairing. 这套设备需要修理。

This phenomenon is worth mentioning. 这一现象值得一提。

Section 2 Quantifying Siting Difficulty: A Case Study of US Transmission Line Siting

Recent decades have seen a growing worldwide demand for new energy infrastructures, including power plants, wind farms, electric transmission lines, liquefied natural gas terminals, and petroleum refineries, among other major projects. Siting such energy facilities, however, has become increasingly difficult. Because of their large scale and technical complexity, many projects involve disparate risks, costs, and benefits for stakeholders, affected populations, and surrounding environments. This asymmetric distribution of project impacts has often fueled intense local opposition and compounded already complex engineering and economic considerations and project constraints.

Siting difficulty is now frequently associated with the familiar acronym NIMBY (not in my backyard) and even more extreme acronyms like BANANA (build absolutely nothing anywhere near anything); however, the problem as a whole is more complex than these expressions suggest. The term siting difficulty, as used here, is defined as any combination of obstacles in facilities planning and siting processes, including public opposition; environmental, topographic, and geographic constraints; interagency coordination problems; and local, state, and federal regulatory barriers to permitting, investment, and/or construction. Siting difficulty is thus a broad and complex problem, affecting a variety of industries, for which solutions are not obvious or well understood.

The lack of substantial data is another major obstacle to understanding the problem. Most academic research and industry trade publications focus on either individual causes of siting difficulty, such as public opposition, or localized effects, such as transmission grid congestion. These analyses are advanced in the absence of any clear empirical reference level for difficulty as a whole, and as a result, many of these studies have limited practical application and policy relevance.

To bridge that gap, this paper develops a policy-level framework for assessing siting difficulty, based on several datasets and statistical analyses. The next section outlines our approach and methods and organizes the sections to follow.

The analytical approach developed in this paper is based on a two-step structure. The first step focuses on answering the question "How difficult is siting?" using a collection of siting indicators. The second step then builds on the resulting measure of siting difficulty to address the question "What makes siting difficult?"

Our formulation is similar to that of current climate change research, where some researchers are looking for "indicators" to determine whether climate change is happening, where it is taking place, and to what extent; and others are examining possible contributing causes and mitigation strategies. Until the significance of the change has been robustly characterized, evaluations of contributing causes (and their interactions) remain out of context.

Similarly, for facilities siting, a quantitative measure of difficulty must first be created and verified, and only then can the causes of siting difficulty be analyzed in context.

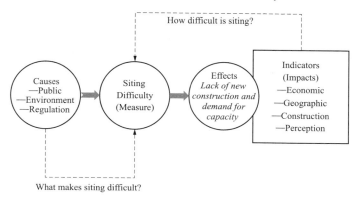

Fig.4.1　Diagram of causes, effects, and indicators of siting difficulty

Fig.4.1 diagrams our framework and highlights the general relationships among our selected siting indicators and the typical causes and effects of siting problems for the case of electric transmission line siting. This diagram illustrates how multiple causes of siting difficulty, such as public opposition, environmental barriers, and regulatory roadblocks, could collectively lead to an underinvestment in infrastructure. The resulting lack of capacity then triggers industry-level economic, physical, and perceptual impacts, such as variations in the cost of electricity generation and changes in capacity additions. These types of large-scale impacts form the basis for the siting indicators in the analyses to follow.

The four indicators in Fig.4.1 are neither direct causes nor effects. Because of the numerous feedback loops and interactions among the causes and effects of siting difficulty, no single cause or effect adequately represents the overall problem. For example, one possible measure of transmission line siting difficulty is the difference between generation and transmission capacity additions; however, this metric could conceivably mask underinvestment in both generation and transmission caused by shared siting constraints. As a result, siting difficulty needs to be quantified based on a careful evaluation and aggregation of multiple impacts.

 原文翻译

第二节　量化选址困难：美国的一个输电线路选址案例研究

近几十年来，包括火力发电厂、风力发电厂、输电线路、液化天然气终端和石油炼油厂等重大项目在内的全球新能源基础设施的需求日益增长。然而这类能源设施的选址变得越来越困难。这是由于该类设施规模较大且工艺较为复杂，许多项目还要考虑不同的风险、成本、利益相关者的收益、受影响的人群和周围环境。这种项目影响程度的不同常常会加剧当地反

对派的紧张情绪，增加已有的工程难度、预算成本和项目限制。

现如今选址问题经常与邻避效应（不要建设在我家后院），甚至是更极端的香蕉效应（不要在其附近建设任何东西）相联系；然而，总的来说这个问题比上述更加复杂。这里所说的选址问题，是指在设施筹备和选址过程中的一切障碍，包括公众反对；环境、地形和地理限制；跨部门的协调问题；当地、州和联邦监管部门在准许、投资以及/或者建设中的阻碍。选址问题十分宽泛和复杂，其影响涵盖各行各业，但该问题还未出现很好的解决办法。

解决选址问题的另一个主要障碍是缺乏大量的数据。大多数学术研究和工商业出版物着重于选址困难的某个原因，例如，公众反对或者如输电网络拥堵这类局部效应。现有分析没有从整体上判断选址困难大小的明确且实证过的参考标准，因此许多研究的实际应用价值和政策相关性有一定局限。

为了解决这个问题，本文基于数个对象和统计分析，从政策层面评估选址的困难。下一节简要介绍了本文所采用的方法，章节组成如下所示。

本文的分析方法包含两步。第一步利用一系列选址指标着重解决"选址有多难"的问题；第二步基于导致选址困难的原因来解决"是什么导致选址困难"的问题。

本文的研究思路和当前对气候变化的研究类似，一部分研究人员寻找"指标"来确定气候变化是否发生，在哪儿发生，发展到什么程度；另一部分研究人员研究气候变化的可能诱因和缓解策略。在气候变化的重要性被揭示之前，其诱因的评定（及其相互作用）一直是片面的。同样的，在设施选址时，必须先对困难程度进行定量检测且对结果进行验证，然后才能分析选址困难的原因。

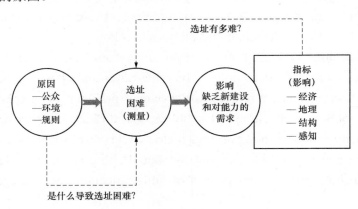

图 4.1　选址困难的原因、影响和指标示意图

图 4.1 展示了输电线路选址问题的整体研究思路，并指明输电线路选址困难的指标、典型原因和其影响之间的关系。该图展现了选址困难的多个原因，如公众反对、环境限制和监管阻碍等，这些都可能导致基建的投资不足。输电容量不足会引发行业级经济上、物质上和感知上的影响，如发电成本和附加投资发生变化。这些类型的大范围影响可作为下文选址困难指标的设定依据。

图 4.1 中的四项指标既不是直接原因也不是直接结果。因为选址困难的原因和结果是由多个反馈和相互作用产生的，没有一个原因或结果充分代表了整个问题。例如，一个有可能评估输电线路选址困难大小的指标是通过比较发电和输电附加容量的大小；然而可以想到的

是,这个指标忽略了在发电和输电中都可能存在的由共享地址造成的投资不足。因此,选址问题需要在仔细评估和综合多种影响的基础上进行量化。

 New Words and Expressions

transmission grid	输电网
wind farm	风电场
current	*n.* 电流
trigger	*vt.* 引发,引起;触发
electricity generation	发电
transmission capacity	输电能力,输电量

 Notes

1. Recent decades have seen a growing worldwide demand for new energy infrastructures, including power plants, wind farms, electric transmission lines, liquefied natural gas terminals, and petroleum refineries, among other major projects.

近几十年全球新能源基础设备的需求日益增长,包括发电厂、风力发电厂、电力输电线路、液化天然气终端和石油炼油厂等重大项目。

2. Siting difficulty is now frequently associated with the familiar acronym NIMBY (not in my backyard) and even more extreme acronyms like BANANA (build absolutely nothing anywhere near anything); however, the problem as a whole is more complex than these expressions suggest.

现如今选址问题经常与邻避效应(不要建设在我家后院),甚至是更极端的香蕉效应(不要在其附近建设任何东西)相联系;然而,总的来说这个问题比上述更加复杂。

3. The term siting difficulty, as used here, is defined as any combination of obstacles in facilities planning and siting processes, including public opposition; environmental, topographic, and geographic constraints; interagency coordination problems; and local, state, and federal regulatory barriers to permitting, investment, and/or construction.

这里所说的选址问题,是指设施在规划和选址过程中的一切障碍,包括公众反对,环境、地形和地理限制,跨部门的协调问题,当地、州和联邦监管部门在准许、投资以及/或者建设中的阻碍。

4. These analyses are advanced in the absence of any clear empirical reference level for difficulty as a whole, and as a result, many of these studies have limited practical application and policy relevance.

现有分析没有从整体上判断选址困难大小的明确且实证过的参考标准,因此许多研究的实际应用价值和政策相关性有一定局限。

5. Our formulation is similar to that of current climate change research, where some researchers are looking for "indicators" to determine whether climate change is happening,

where it is taking place, and to what extent; and others are examining possible contributing causes and mitigation strategies.

本文的研究思路和当前对气候变化的研究类似，一部分研究人员寻找"指标"来确定气候变化是否发生，在哪儿发生，发展到什么程度；另一部分研究人员研究气候变化的可能诱因和缓解策略。

6. Because of the numerous feedback loops and interactions among the causes and effects of siting difficulty, no single cause or effect adequately represents the overall problem.

因为选址困难的原因和结果是由多个反馈和相互作用产生的，没有一个原因或结果充分代表了整个问题。

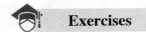

Exercises

Ⅰ. Choose the best answer into the blank.

1. Recent decades have seen a growing worldwide demand for new energy infrastructures, including power plants, ____ , electric transmission lines, liquefied natural gas terminals, and petroleum refineries, among other major projects.

  A. nuclear power plant   B. hydraulic power plant
  C. wind farms   D. thermal power plant

2. The lack of substantial ____ is another major obstacle to understanding the problem.

  A. date   B. experiments   C. theories   D. practices

3. The analytical approach developed in this paper is based on a ____ structure.

  A. four-step   B. three-step   C. one-step   D. two-step

4. The ____ indicators in Fig. 4.1 are neither direct causes nor effects.

  A. three   B. two   C. four   D. five

Ⅱ. Answer the following questions according to the text.

1. What do demand for new energy infrastructures include?
2. What does NIMBY mean?
3. What does BANANA mean?
4. What is the first step of the analytical approach developed in this paper?
5. How many indicators are suggested in Fig. 4.1? Try to list all indicators.

Ⅲ. Translate the following into Chinese.

Siting difficulty and its associated constraints are not monolithic. This paper also makes a first step toward breaking down the causes of siting problems into manageable pieces for evaluation and planning, while simultaneously maintaining a large-scale view of the problem. The results here are not intended to identify and blacklist areas of high siting difficulty or to suggest that all siting difficulty can be predicted and addressed in advance of a planning process. Nor are our analyses the only appropriate characterizations of a broad and complex problem. This work is simply intended to give structure to the ever expanding discussion of energy facilities siting, management, and planning.

As more parties have become involved in the debate over siting, technical solutions and policy solutions to infrastructure demand and siting difficulty have increasingly diverged. Successful development of energy infrastructures requires the integration of both technological system-level innovations and large-scale policy changes. This paper serves as an initial bridge between the quantitative and qualitative issues affecting siting, where a sound strategy for managing siting problems is critical to the success of many energy industries.

Keys

Ⅰ. Choose the best answer into the blank.
1. C   2. A   3. D   4. C

Ⅱ. Answer the following questions according to the text.
1. Power plants, wind farms, electric transmission lines, liquefied natural gas terminals, and petroleum refineries.
2. Not in my backyard. It means they support the project, but oppose the construction near their house.
3. Build absolutely nothing anywhere near anything. It means they oppose the construction near anybody.
4. The first step focuses on answering the question "How difficult is siting?" using a collection of siting indicators.
5. Four. Economic, geographic, construction, perception.

Ⅲ. Translate the following into Chinese.

选址困难和其相关限制不适合从整体上考量。本文将大规模的选址问题分解成各个便于评估和规划的小模块。本文所得结果不是为了识别选址难度高的区域，也不是为了在规划进程之前预测和处理所有的选址问题。本文仅是对广泛而复杂的选址问题进行合理的特性分析。该研究仅为能源设施选址、管理和规划提供更多参考。

随着越来越多的政党参与选址问题的讨论，关于基础设施需求和选址问题的技术方案与相关政策日益分化。较好的发展能源基础建设需集合技术体制创新和大规模政治改革。许多能源产业成功的关键在于有一个高效彻底的策略用于管理选址问题，而本文的研究成果首次将选址困难定量的与定性的影响连接起来。

翻译技巧之被动句（二）

被动句翻译时按译成汉语被动句和译成汉语无主句的方法介绍如下：
1. 译成汉语被动句
（1）译成明显的"被"字句，通常带有"be+动作发出者"或"be called"。
The number of TTL devices that one TTL output will drive is called fanout. 一个 TTL 输出所能驱动的 TTL 装置的数量被称为输出能力。

Gears are used in place of be drivers and other forms of friction drivers. 齿轮(传动)被用来代替带传动及其他的一些摩擦传动形势。

In this case,the molecule is polarized by the field. 在这种情况下，该分子被场极化了。

（2）译成"由、由……组成，用、用于……"等，通常带有"be made up of""be composed of""be used in/to"。

The magnetic field is produced by an electric current. 磁场由电流产生。

Usually alloy is composed of a base metal (the largest part of the alloy) and a smaller amount of other metals. 通常，合金由一种贱金属（合金中最多的那部分）和少量的其他金属组成。

They are used to support and position a shaft and to reduce the friction created by the rotating part,particularly when under load. 它们用于轴的支承、定位，并降低由转动件带来的摩擦，特别是在载荷作用下。

（3）译成判断句式"是……的"。如果英语句子本身所强调的是与某一静态动作有关的具体情况，如时间、地点、方式、方法等；或着重说明一件事情是"怎样做""什么时候、什么地点做"的等，这时我们就可以把这些情况放在"是……的"这种判断句式中，使之突出。这种判断句式能够清楚地表达作者的意图和客观事实。

Everything in the world is built up from atoms. 世间万物都是由原子构成的。

The AIDS virus was found in human white blood cells in 1983. 艾滋病病毒是1983年在人体白血球内发现的。

Iron is extracted form the ore of the blast fumace. 铁是用高炉从铁矿中提炼出来的。

（4）在谓语动词前面添加"予以""加以""受到""得以"等词。

The translation technique should be paid enough attention to. 翻译技巧应予以足够的重视。

Other mistakes of this dictionary will be corrected in the next edition. 这本辞典的其他错误将在下一版中予以修正。

Coal and oil are the remains of plants and animals. Crude mineral ores and crude oil must be purified before they can be used. 煤和石油是动植物的残骸。原矿石和原油必须加以精炼才能使用。

The temperature must be controlled to produce the desired qualities in the steel. 为了得到（人们）想要的性能，热处理时钢的温度必须加以控制。

Technology have been rapidly developed because of the discovery of electricity. 由于电的发现，科学技术得以迅速发展。

Teachers should be respected. 教师应受到尊重。

2. 译成汉语无主句

当无法知道或无法说出动作发出者时，往往可以把英语的被动语态译成汉语的无主句。

To get all the stages of the ground,a first big push is needed. 为了使火箭各级全部离开地面，需要有一个巨大的第一推动力。

Work is done,when an object is lifted 当举起一个物体时，就做了功。

Attention must be paid to safety in handling radio active materials. 处理放射性材料时必

须注意安全。

总之，英语中被动语态的翻译不能一概采用所谓"语言等值"的顺译法，而必须根据汉语的语法和习惯，发挥汉语的优势，用规范化的汉语表达方式，忠实而恰当地反映出原作语言的真实含义，使译文的形式与原文内容辩证地统一起来，才能收到良好的翻译效果。

Section 3 Reliability–Based Transmission Line Design

It is well known that environmental loads, e.g., wind and ice, acting on power transmission lines are highly uncertain, as are the structural strengths of the towers supporting the lines. The design of such systems must take uncertainty into account in order to achieve acceptable reliability at a reasonable cost. The paper presents a simulation-based methodology for the optimal design of a transmission line which considers uncertainties in both environmental loads and structural resistance. The methodology is developed and illustrated for the simple problem of determining the optimal span length required for designing against tower failure. Wind, ice and tower resistances are simulated over the extent of the transmission line and over the design life of the transmission system. Total expected system cost, along with the estimated probability of lifetime failure, are produced for a range of possible span lengths, allowing an informed decision regarding the optimum span length for the tower strength limit state.

The goal of a reliability-based design is to produce a system which is both robust and economical. In order to achieve this goal, design decisions must be optimally made in the face of uncertainty. Although uncertainty can be reduced to some extent through experimentation and statistical studies, it remains significant due to variability in loads, construction quality, model/design error, and system degradation. In particular, uncertainty in environmental loading due to extreme climatic events has lead to a re-evaluation of current transmission line design and upgrading practices.

This paper presents a reliability-based methodology for the design of an electrical power transmission system consisting of conductors and transmission towers spanning between a generating plant and a destination group of customers (e.g., a small city, or a segment of a large city). To develop and illustrate the methodology, a single design decision will be examined—optimization of the conductor span between towers with respect to the tower failure limit state. Deciding on a span length depends on many possible limit states (e.g., tower capacity, sag, vibration, and tension). In this paper, only the limit state of tower capacity will be considered, although it is recognized that this limit state may not govern the design and a fully reliability-based design must apply the same ideas expressed in this paper to all other limit states and include all other design variables. As such, it is emphasized that the proposed methodology is to be viewed as a supplement to current design methods (since the latter consider all limit states and design variables) and as an increment to the pioneering work of other reliability-based power transmission researchers (e.g., Ghannoum and Phoon).

Regarding the tower capacity limit state, as the span between towers increases, the conductor, ice and wind loads acting upon each tower increases, resulting in an increased probability of tower failure for a given tower design. On the other hand, increasing the span reduces the required number of towers and thus the initial system cost. The best design span with respect to the tower capacity limit state will involve a tradeoff between the cost of failure over the design life of the transmission system and the initial system cost.

A reliability-based design considers both the probability of an adverse event occurring and the consequence of that event should it occur to determine an optimal design decision. Consequence is generally expressed in monetary terms and typically based on the cost of repairs, remediation, human safety, inconvenience, or other losses. The structural failure of a transmission tower is an event with multiple adverse consequences including possible power outages, tower replacement costs, incremental generation costs (running a generator or purchase from another source), and other incidental costs which are difficult to quantify, such as unfavorable consumer perception. Power outages, in turn, can lead to losses associated with homes (e.g., refrigerator contents), businesses, and even personal safety, particularly during the winter.

When new transmission lines are designed, or older established lines reassessed, it is desirable to minimize costs while maintaining an acceptably low probability of failure. Electrical power providers are beginning to recognize the value of reliability-based design. For example, Hydro Quebec has been using probability based design techniques for new tower designs and upgrade of older transmission lines for many years now. This task involves minimizing the sum of initial capital and transmission line failure costs while maintaining an acceptable reliability. Ghannoum suggests that the optimum annual transmission line failure probability should range from about 0.01 to 0.001, depending on the consequences of failure. Acceptable failure probabilities in this range correspond to climatic loads having return periods of roughly 50 to 500 years.

This paper will develop a reliability-based design methodology aimed at minimizing the total expected cost of a transmission tower system over a design life of 50 years. The design decision variable considered will be the conductor span between transmission towers, assuming level terrain, and the decision will be made on the basis of climatic loads as they affect the possibility of tower structural failure. The case study is hypothetical, and only considers one of myriad design decisions and a single decision criterion, but serves to demonstrate a simulation approach to reliability-based design.

In contrast to traditional design, a reliability-based approach needs to identify the random quantities on both the load and resistance side, and to gather enough data to allow the distributions of these random quantities to be estimated. In addition, because loads and strengths are both time varying and because maintenance/failure costs accrue with time, the system lifetime needs to be carefully considered. In other words, while traditional designs typically need only mean (or characteristic) load and resistance parameters, along with empirically based safety factors, a reliability-based approach also needs to know about;

construction, maintenance, and failure costs; the complete distribution of all random load and resistance parameters; and how all of these parameters vary with time. This represents quite a bit of additionally required information, much of which is not currently known. It is anticipated that, as reliability-based design tools such as proposed in this paper become readily available, the data-base required to estimate distributions will become available in the years to come.

原文翻译

第三节　基于可靠性的输电线路设计

众所周知，环境荷载对输电线路的作用是非常不确定的，如风和覆冰、支撑导线的杆塔的结构强度也是如此。这类输电系统的设计必须将这种不确定性考虑在内，从而以合理的成本获得可接受范围内的可靠性。本文同时考虑了环境荷载和结构承载力，提出了一种对输电线路进行最优化设计的仿真方法。这种方法将通过确定最优挡距来避免杆塔出现故障的简单问题来阐述和验证。风荷载、覆冰荷载和杆塔承载力在仿真过程中会超出输电线路能够承受的极限范围，而且也会超过输电线路的设计使用寿命。不同挡距具有不同的总预期成本和在使用寿命中发生故障的概率，这说明在杆塔强度的极限范围内有一个最优的挡距。

基于可靠性设计的目的是为了设计出一个既可靠又经济的输电系统。为了达到这个目的，在做设计决策的时候就必须要面对这些不确定性。尽管这些不确定性能够通过实验和统计研究的方式减少到某种程度，但仍然会留有大量的不确定性，这是因为荷载、工程质量、模型/设计误差和系统老化具有多样性。特别的，由极端气象引起的环境荷载的不确定性已经导致了现存的输电线路设计和升级改造的标准需要进行重新设定。

本文介绍了一种基于可靠性的方法对输电系统进行设计，包括导线和发电厂与大量客户（例如，一个小城市或者大城市的一部分）之间的杆塔挡距的设计。为了阐述和证实这个方法，本文将会提出一个只考虑单个因素的设计方案——在多基杆塔中的一基达到极限状态情况下对挡距进行最优化。确定挡距需要依据较多可能的极限状态（比如杆塔承载力、弧垂、振动和张力）。本文只考虑杆塔承载力的极限，尽管杆塔承载力的极限在设计中并不是最重要的，但一个完整的基于可靠性的设计可以采用相同的思想来处理所有其他的极限状态，以及其他的所有设计参量。同样的，需要强调的是本文提出的方法可以作为现行设计方法的补充（由于本文的方法考虑了所有的极限状态和设计参数），并且可以促进其他基于可靠性的输电线路研究的创新性工作（比如格安诺姆和弗恩）。

当杆塔处于承载力极限状态时，随着挡距的增加，导线、覆冰和风荷载对每一基的杆塔所起的作用也相应增大，这会使已设计的杆塔出现故障的概率增大。另一方面，增大挡距可以减少所需的杆塔数量和一次系统投资。对于杆塔承载力极限状态下最优的设计挡距需要考虑杆塔故障所需费用、输电系统的设计寿命和系统一次投资。

基于可靠性设计的方法也要考虑不良事件发生的概率及其带来的结果来确定设计方案。其结果一般用货币的方式来表达，典型的就是根据维护、修理、人身安全、不便或其他损失的费用。输电线路杆塔结构失效是一个具有多种不利后果的事件，这些不利后果包含可能导

致电力中断、杆塔更换费用和发电成本增加（多启动一个发电机或从另一个电源买电），还有其他难以量化的附加成本，比如令用户不愉快的感觉。反过来，电力中断将导致和家庭有关的损失（如冰箱里的物资）、商业，甚至是个人安全，特别是冬季。

当新的输电线路进行设计或老旧线路需要重新设计时，保证较低的故障率并能最小化成本是非常令人期待的。电力供应商开始意识到基于可靠性的输电线路设计的价值。例如，魁北克水电公司至今已经多年采用基于可靠性对新的杆塔进行设计，并且对老旧输电线路进行更新。这项任务主要是将一次总投资和保持允许范围内的可靠性所产生的输电线路损失费用最小化。格安诺姆依据故障所造成的损失，建议最适合的输电线路故障概率应该为 0.001～0.01。在这个范围内允许的故障概率大概相当于具有 50～500 年重现一次的气象荷载出现的概率。

本文将开发一种基于可靠性的设计方法，旨在将设计寿命为 50 年的输电杆塔系统的总预计成本降至最低。对于平坦地区，以导致杆塔结构失稳的气象荷载为根据，所考虑的各种变量中最关键的则是两个杆塔之间的导线或档距。案例研究提出了一种假想，即只考虑众多设计变量中的一种，并且只制定一个决策标准，但该标准可以用来验证基于可靠性设计的仿真方法的正确性。

与传统设计形成对比，基于可靠性设计的方法需要在荷载和承载力两方面定义随机量，同时也需要足够的数据用来预估这些随机量的分布规律。此外，由于荷载和强度都会随着时间变化以及维修与损失费用随着时间增加，因此需要仔细斟酌这些因素对系统寿命的影响。换句话说，当传统的设计一般只需要平均（或者典型）荷载和承载力参数以及根据经验确定的安全系数时，基于可靠性的设计方法也需要知道这些；建造、维护和故障损失费用；随机荷载和承载力参数完整的分布；和这些参数是怎样随着时间发生变化的。这意味着需要相当多的额外必要资料，而其中一部分到现在还不能得到。可以预料的是，如本文提到的基于可靠性设计工具如果能够实现，那么预估分布规律的数据库将在不久的将来可以使用。

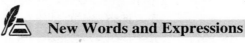

New Words and Expressions

sag	*n.* 松弛
multiple	*n.* 并联
electrical power	电源；电功率
generating plant	发电站；发电厂；发电设备
load	*n.* 负载，负荷

Notes

1. Although uncertainty can be reduced to some extent through experimentation and statistical studies, it remains significant due to variability in loads, construction quality, model/design error, and system degradation.

尽管这些不确定性能够通过实验和统计研究的方式减少到某种程度，但仍然会留有大量的不确定性，这是因为荷载、工程质量、模型/设计误差和系统老化具有多样性。

2. Regarding the tower capacity limit state, as the span between towers increases, the conductor, ice and wind loads acting upon each tower increases, resulting in an increased probability of tower failure for a given tower design.

当杆塔处于承载力极限状态时，随着挡距的增加，导线、覆冰和风荷载对每一基的杆塔所起的作用也相应增大，这会使已设计的杆塔出现故障的概率增大。

3. When new transmission lines are designed, or older established lines reassessed, it is desirable to minimize costs while maintaining an acceptably low probability of failure.

当新的输电线路进行设计或老旧线路需要重新设计时，保证较低的故障率并能最小化成本是非常令人期待的。

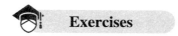

Exercises

Ⅰ. Choose the best answer into the blank.

1. The goal of a reliability-based design is to produce a system which is both robust and _____.
    A. economical　　　B. secure　　　C. stable　　　D. time varying

2. This paper presents a reliability-based methodology for the design of an electrical power transmission system consisting of conductors and transmission towers spanning between a/(an) _____ and a destination group of customers.
    A. user　　　B. dispatcher　　　C. operator　　　D. generating plant

3. Acceptable failure probabilities in this range correspond to climatic loads having return periods of roughly _____ years.
    A. 20 to 30　　　B. 30 to 50　　　C. 50 to 500　　　D. 500 to 700

4. Increasing the span _____ the required number of towers and thus the initial system cost.
    A. reduces　　　B. increases　　　C. enlarge　　　D. shrink

Ⅱ. Answer the following questions according to the text.

1. Why does the uncertainty remain significant?
2. Deciding on a span length depends on many possible limit states. Try to list a few limit states.
3. Which limit state is considered in this paper?
4. What can result in an increased probability of tower failure?
5. What are the adverse consequences of the structural failure of a transmission tower?

Ⅲ. Translate the following into Chinese.

The methodology developed here forms just one tool in a transmission line designer's repertoire. The current approach to transmission line design involves the following steps: ① mapping out the terrain and determining environmental loads; ② locating the tower positions according to the terrain, loads, maximum sag, and other conductor demands; and

③ selection of suspension and dead-end towers to safely support the conductor loads. The last step would typically be done using finite element analysis. The results of this paper can be used to aid in step ③, i.e., in the optimal reliability-based selection of towers. If the methodology proposed here is applied to a variety of tower designs to obtain an optimal span for each, then a table of tower types versus optimal spans could be developed. Towers having optimal spans closest to the spans required by step ② could then be selected.

Keys

Ⅰ. Choose the best answer into the blank.
1. A 2. D 3. C 4. A

Ⅱ. Answer the following questions according to the text.

1. It remains significant due to variability in loads, construction quality, model/design error, and system degradation.

2. For example, tower capacity, sag, vibration, and tension.

3. Tower capacity.

4. As the span between towers increases, the conductor, ice and wind loads acting upon each tower increases, resulting in an increased probability of tower failure.

5. The structural failure of a transmission tower is an event with multiple adverse consequences including possible power outages, tower replacement costs, incremental generation costs, and other incidental costs.

Ⅲ. Translate the following into Chinese.

本文的方法只是作为输电线路设计师电脑中的一个可以使用的工具。目前输电线路设计方法包括以下几个步骤：①标注地形并确定环境荷载；②根据地形、荷载、最大弧垂和其他导线需求确定杆塔位置；③选择可靠支撑导线的耐张塔和终端塔。最后一步通常利用有限元分析解决。本文结论可用于帮助解决步骤③，即基于可靠性对杆塔进行优化选择。如果采用本文所提的方法为各种杆塔设计提供最优挡距，那么接着可得到不同塔型及其最优挡距的表格。最接近步骤②要求的挡距的杆塔即为所选杆塔。

翻译技巧之被动句（三）

1. 对于英语双重被动句的常见译法

英语双重被动句的译法类似于一般被动句的译法，但英语双重被动句在译成汉语时，不能生搬硬套，而要根据汉语的习惯，采取一些辅助、综合的办法，因为英语中有双重被动句，而汉语中没有相应的句式。

（1）英语双重被动句译成含有泛指主语的汉语主动句，译句前通常加上"人们""大家""有人""我们"等泛指词，原句中的主语与第二个不定式被动结构扩展为一个新的分句。例如：①The river is known to have been polluted.人们知道，这条河已经受了污染。②The meeting is suggested to be put off till next Friday.有人建议会议推迟到下周星期五举行。

（2）英语双重被动句译成汉语无主句。原句中的主语与第二个不定式被动结构扩展成为一个新的分句。例如：①The guests were arranged to be met at the meeting room.安排在会议室见客人。②Cars are allowed to be parked over there.可以把车停在那边。

（3）英语双重被动句在汉译时，有时可以按顺序直接省略被动语言标志的汉语被动句。例如：①The books aren't permitted to be taken out of the room.这些书不许带出房间。②That film is not allowed to be shown.这部影片不许放映了。

（4）有的英语双重被动句在汉译时，可根据第一个被动结构的词汇意义译成"据……"，原句中的主语与第二个不定式被动结构扩展为一个新的分句。例如：①The child is reported to have been found. 据报道孩子已经找到了。②The wedding is supposed to be announced soon. 据估计婚礼不久就宣布。一般来说，英语被动语态是由"be+及物动词的过去分词"构成的，但是这并不是被动语态的唯一表示方式，除了助动词be外，有些动词还可以用来构成被动语态，弄清楚一问题，有利于翻译英语被动句。此外，"动词get、become 或 feel 的一定形式+及物动词过去分词"可表示被动语态，这一结构通常表示动作的结果而不是非动作本身，也常用来表示突然发生、未曾料到的事。因此，这类句子常译成汉语被动句或汉语主动句。

2. 主动语态表示被动意义时的译法

有些英语句子从结构上看是主动语态，但实际上是表达动作意义，可看成隐形被动语态。翻译时常译成主动语态要有如下几种情况：

(1) Sth needs (want, require, demand…) doing Sth.

The machine needs repairing. (=The machine needs to paired.) 这机器需要修理。

(2) Sth be worth doing

The book is worth reading. (=The book is worthy to read.) 这本书值得看一看。

(3) S.b has (have) sth to do

He has a lot of work to do. (=A lot of work will be done by him.) 他要做许多工作。

(4) Sth be building (printing, cooking…)

The factory is building. (=The factory is being built.) 工厂正在修建中。

(5) Sth sells (weigh, wash, record, taste…).

The machines made in China sell very well abroad. 中国制造的机器在国外畅销。

Chapter 5 Construction Technology

Section 1 Overview of the Transmission Line Construction Process

This paper outlines various procedures followed in the construction of transmission lines designed by Los Angeles Department of Water and Power (LADWP). The transmission line network designed or constructed by LADWP spans about 3500 circuit miles with voltages ranging from 115kV AC to 1000kV DC. Since most lines constructed by LADWP in recent years have been extra-high-voltage AC, this paper emphasizes the construction of 500kV AC transmission lines built in the southwest United States. The paper presents some concepts and empirically developed practices that have proven successful over many years of transmission line construction by LADWP.

1. Introduction

Various aspects of the construction process used by Los Angeles Department of Water and Power (LADWP) in constructing extra-high-voltage AC transmission lines are presented in this paper. The construction process addresses such activities as tower footings, conductor installation, clipping offsets, hold-down weights, jumper cables, vibration dampers, spacer installation, insulators, splicing, fence grounds, radial counterpoises, expenditure and construction materials.

2. Tower footings

Tower footings are designed by the structural design engineer. The design is based on the following design data.

(1) Tower design (e.g. type, loading).

(2) Basis of soil allowables (e.g. allowable lateral bearing stresses and skin friction for drilled piers, the allowable bearing stress and soil densities for embedded pads, and the capacities of rock bolts of various embedment lengths) and soil corrosives (high sulfate and chloride content in soils has a detrimental effect on concrete unless the concrete is designed for it).

(3) Foundation selection. Drilled piers are the preferred alternative whenever possible in terms of economics and reliability. Embedded pads are used for soils that cannot be augered (such as areas heavy in boulders or in areas of collapsing sand). For foundations that are located in bedrock, rock-type footings are used. Grillage foundations are used in areas where roadless construction is required.

(4) Grounding requirements.

(5) Curb height. LADWP's minimum allowable curb height is 6 in. (15.24cm) above

ground level. The maximum allowable curb height is 4 ft (122cm).

3. Installation of hold-down weights

Ideally, the only time weights should be used is as a jumper restraint on dead-end towers. However, in practice hold-down weights have found two basic applications. Both apply specifically to suspension towers with uplift problems, and both serve to ensure the tension on an insulator string is adequate to prevent corona or interference problems (minimum weight on a 500kV line's insulator string is 500lb).

First, hold-down weights are used in a hilly area on a tower that is low enough to adjacent towers to have a marginal uplift problem. Uplift causes corona and interference problems because of the incorrect tension on the ball-and-socket joint of the insulator strings. The solution for any uplift problem is to install a larger tower. However, when the line is under construction the logistics and cost of specially ordering a larger tower may be prohibitive. In such a case hold-down weights are used to correct the uplift and subsequent insulator tensioning problem.

Second, hold-down weights are used in windy areas for spans having large (wind span)/(weight span) ratios. Large ratios occur with long spans and in a windy area can lead to uplift problems. The wind span is the length of two adjacent spans divided by two. The weight span is the distance between two low points of the catenaries of adjacent spans. It is evident that installing a larger tower will increase the distance between catenary low points and decrease the ratio (this is the preferred solution). Installing hold-down weights will not substantially affect the ratio, but will artifically increase the tension on the insulator strings and solve the corona problems.

Typically, hold-down weights range from 300 to 5000 lb (136~2268kg). The use of hold-down weights is not recommended if weights larger than 5000lb are required.

4. Installation of jumper cables

Jumper cables are used to connect conductors across dead-end towers. The calculation of jumper cable lengths varies, with each utility developing its own method empirically. Theoretical models are difficult because jumper cables are too small to approach a catenary or parabolic shape. Therefore computer models or graphical methods are employed to determine jumper lengths. A field method used by contractors is to climb the tower, simulate the jumper by using a rope, gage the correct length by measuring clearances achieved, then cut the conductor to match the rope length. Not very sophisticated, but it produces acceptable results.

5. Conductor splicing

(1) All splices should be located in the span at least 50ft (15.2m) away from any suspension or dead-end hardware. No conductor in a span should have more than one splice, and no splice should be made in any span crossing a railroad, major telephone line, state or federal highway, or power line of 66 kV or above.

(2) Splices should be shaped and finished after compression to produce a smooth surface

free of flash or sharp points which might be a source of corona or radio interference. All splices should be straight after installation.

6. Installation of danger signs

(1) Danger signs reading "Danger" "High Voltage" and "Keep Off" should be posted on the right-hand part of the four faces of all tower pedestals. These signs should be placed on the first horizontal members at least 7.5ft (2.1m) above the footings.

(2) Aerial patrol mile markers and crossing signs should be posted on signs outside the city limits likely to be patrolled by aircraft.

7. Installation of fence grounds

(1) Fences in the vicinity of the right of way receive lines of flux from the transmission line which may cause hazardous shocks. To reduce the likelihood of injury, the wires of the fence are connected together with #12 galvanized iron wire. This wire is then connected to a 0.625 in. (1.59cm) diameter Copper weld ground rod, 8ft (2.4m) in length, which is inserted approximately 7.7ft (2.3m) into the ground. In the southwest United States the rod must be driven to this level to reach the permanent moisture level of the soil.

(2) The Copperweld ground rod is a steel rod sintered (a type of powder metallurgy) with a thin copper coating that increases the grounding effect. Very little change in resistance would result from using larger diameter ground rods, as it is mainly the soil surrounding the electrode and not the diameter of the rod that determines the resistance.

8. Miscellaneous construction activities

(1) Construction of access roads.

(2) Construction of culverts and drains.

9. Construction materials

(1) Material furnished by LADWP: footing stubs, towers, tower and step bolts, conductors, ground wires, electrode line conductors, insulators, hardware, accessories.

(2) Material furnished by the construction contractor: culverts, drains, gates, concrete, reinforcing bars, cattle guards, fence grounds, counterpoises and clamps.

10. Expenditure

(1) The cost of designing and constructing a transmission line depends on many factors. Terrain, location, distance traversed, conductor configuration and new environmental regulations are a few of the typical factors which vary from line to line and change its associated costs.

(2) The cost of a 200 mile (322km), single-circuit, 500kV AC, three-conductor bundle passing through mostly federal land in the southwest US is approximately $700 000 per mile ($435 000 per km). These costs are for a line being built in the 1992~1995 timeframe and do not include the cost of station structures.

(3) A two-conductor bundle traversing the same area would average approximately $500 000 per mile ($311 000 per km). The extra power carried by the three-conductor bundle would more than offset the extra cost incurred during construction.

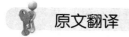

第一节　输电线路建设过程概述

本文概述了洛杉矶水电部（LADWP）在建设输电线路过程中的所有内容。洛杉矶水电部设计并建设完成的输电线路总长约 3500 英里，涵盖了从 115kV 交流电压到 1000kV 直流电压的所有电压等级。近年来，洛杉矶水电部建设的多是超高压交流输电线路，因此，本文主要分析了建造在美国南部的 500kV 交流输电线路，并提出了一些概念和经验。这些经验在洛杉矶水电部多年的输电线路架设中被证实是成功的。

1. 引言

本文全方位地阐述了洛杉矶水利电力部门施工建设特高压交流输电线路的全过程。该施工过程主要涉及杆塔基础施工，导线架设，塔材剪切，重物压紧，跳线、减振器、间隔棒、绝缘子的安装，导线压接，接地网，径向平衡重锤，耗材和建筑材料等一系列内容。

2. 杆塔基础

结构设计工程师在设计杆塔基础时需要考虑如下设计数据。

（1）杆塔基础结构（如类型和荷载）。

（2）土壤的基本选择依据（如钻孔时桩式基础的允许侧向承载应力和表面摩擦力，嵌入式基础的允许承载应力和土壤密度，以及各种嵌入长度下的石质锚桩容量）和土壤腐蚀特性（除非对混凝土进行特殊处理，否则土壤中含量较高的硫酸根和氯离子都会对混凝土产生不利影响）。

（3）杆塔基础选型。出于对经济性和可靠性的考量，钻孔桩式基础是首选，而嵌入式基础则可用于不能钻孔的土地（如巨石区或塌沙区）。位于岩石地基的基础需用石质类材料，而对于没有路的地区，则需要用装配式基础。

（4）接地要求。

（5）高度限制。洛杉矶水电部允许最低控制高度为 6 英寸（15.24cm），允许最大控制高度为 4 英尺（122cm）。

3. 压紧重锤

理想情况下，唯一的一次压紧重锤应用于终端塔跳线的约束。然而实际上，压紧重锤有两类基本用途。这两类用途均适用于有关耐张塔的上拔问题，以及确保绝缘子串张力防电晕、抗干扰的问题（500kV 线路绝缘子串的最低压紧力为 500 磅）。

首先，当将压紧重锤用于丘陵地区的杆塔时，该塔位置须低到足以与邻近的杆塔之间有临界上拔问题。由于绝缘子串球窝式接头存在不正常张力，这种上拔将引起电晕放电和无线电干扰问题。安装一个更大的杆塔是解决一切上拔问题的基本方法。但当架设这类线路时，定制这种杆塔的成本及其运输问题可能让人望而却步。在这类情况下，压紧重锤通常被用来纠正上拔问题和随后的绝缘子张力问题。

其次，压紧重锤可用于大比例挡距（风载挡距，水平挡距）/（重力挡距，垂直挡距）的有风地区。这类大比率挡距发生在大挡距和有风地区时，可能会引起上拔问题。水平挡距为两相邻挡距的一半，而垂直挡距是指相邻挡距悬链线两最低点之间的距离。显然，安装一个较大的杆塔能增加两悬链线最低点之间的距离从而降低比率（这是首选解决方案）。虽然安装重锤不会从本质上影响挡距比率，但能人为地增加绝缘子串上的张力以解决电晕问题。

通常，压紧重锤的范围为 300 到 5000 磅（136~2268kg）。不推荐使用超过 5000 磅的压紧重锤。

4. 跳线的安装

跳线用于连接穿过终端塔的导线，其长度计算方法有多种，这些方法都是凭经验得到的。由于理论模型中跳线太小，接近一个悬链或抛物线形状，运用理论模型计算跳线长度较为困难，因此，通常采用计算机模型或图解方法来得到结果。施工承包商常采用一种现场方法，此方法通过爬上塔架，用绳子模拟跳线，再通过测量绳子来获取正确的跳线长度，最后切割与绳子长度相同的导线，即为所求跳线。该方法过程较为简单，结果也满足工程要求。

5. 导线的压接

（1）任一接头都应与所有悬挂硬件、终端型硬件至少相隔 50 英尺（15.2m）。挡距中的导线不应多于一个接头，并且在穿越铁路、主要的电话线、州或联邦公路，或 66kV 及以上的导线的档距中不应有任何接头。

（2）接头在压紧完成后的形状应该表面光滑，以避免接头上产生任何的毛边或尖点。这些毛边或尖点都可能是电晕源或无线电干扰源，此外，所有接头应安装在一条直线上。

6. 危险标识的安装

（1）危险标识，如"危险""高压"和"远离"应张贴在塔底座四个面的偏右，且高于杆塔基础至少 7.5 英尺（2.1m）的水平位置。

（2）高空航线英里标识和跨越标识应张贴在可能被飞机航行的城市范围以外的标志上。

7. 接地网的安装

（1）右侧的接地网接收来自输电导线的通量，可能会引起有害的冲击。为了减少此类损伤，可以将接地网线与 12 号镀铁电线连接在一起，再将此电线连接到一个直径 0.625 英尺（1.59cm），长 8 英尺（2.4m）的铜钢丝接地杆，该接地杆被插入地面大约 7.7 英尺（2.3m）处。在美国西南部，这种接地杆必须插到该深度才能达到土壤的永久湿度水平。

（2）包铜钢丝接地杆是一根用薄铜涂层烧结的钢筋（一种粉末冶金），可以增加接地效果。较大直径的接地杆有着很小的电阻，其电阻大小由电极周围的土壤决定，而不是由杆的直径决定。

8. 其他建设活动

（1）线路通道的建设。

（2）涵洞及排水沟的建设。

9. 建筑材料

（1）由洛杉矶水电部提供的材料：塔基立柱、杆塔、杆塔和阶梯型螺栓、导线、接地线、电极导线、绝缘子、硬件、配件。

（2）由施工承包商提供的材料：地下管道、排水管、门、混凝土、钢筋、防畜栏、栅栏、接地网、平衡锤和夹具。

10. 财政支出

（1）输电线路设计和施工成本取决于许多因素。地形、位置、穿越距离、导线的配置和新的环境法规等一些典型因素的差异，都会影响线路建设的相关成本。

（2）一条长为 200 英里（322km），单回路，500V 交流，穿越美国大部分西南部联邦土地的三分裂导线，其成本约为 70 万美元（每公里 43.5 万美元）。这个价格可以在 1992～1995 年建设一条不包括变电站结构的线路。

（3）一条两分裂导线穿越同一地区，平均每英里只需花费约 50 万美元（每千米 31.1 美元），但由三分裂导线输送的额外电能足以抵消施工过程中所产生的额外费用。

 New Words and Expressions

AC(alternating current)	交流电
DC(direct current)	直流电
conductor	n. 导体
spacer	n. 逆电流器
insulator	n. 绝缘子
chloride	n. 氯化物
vibration damper	减振器
dead-end tower	终端塔
suspension tower	直线塔
aeolian vibration	风激振动
wind span	水平挡距
weight span	垂直挡距

 Notes

1. A field method used by contractors is to climb the tower, simulate the jumper by using a rope, gage the correct length by measuring clearances achieved, then cut the conductor to match the rope length.

施工承包商常采用一种现场方法，此方法通过爬上塔架，用绳子模拟跳线，再通过测量绳子间隙来获取正确的跳线长度，最后切割与绳子长度相同的导线，即为所求跳线。

2. Theoretical models are difficult because jumper cables are too small to approach a catenary or parabolic shape. Therefore computer models or graphical methods are employed to determine jumper lengths.

由于理论模型中跳线太小，接近一个悬链或抛物线形状，运用理论模型计算跳线长度较为困难，因此，通常采用计算机模型或图解方法来得到结果。

3. All splices should be located in the span at least 50ft (15.2m) away from any suspension or dead-end hardware. No conductor in a span should have more than one splice, and no splice should be made in any span crossing a railroad, major telephone line, state or federal highway, or power line of 66 kV or above.

任一接头都应与所有悬挂硬件、终端型硬件至少相隔 50 英尺（15.2m）。挡距中的导线不应多于一个接头，并且在穿越铁路、主要的电话线、州或联邦公路，或 66kV 及以上的导线的挡距中不应有任何接头。

Exercises

I. Choose the best answer into the blank.

1. How many design data the structural design engineer should be followed? ____
   A. 2　　　　　　　B. 3　　　　　　　C. 4　　　　　　　D. 5
2. Jumper cables are used to connect ____ across dead-end towers.
   A. conductors　　　B. insulators　　　C. transmission lines　D. electrodes
3. Which danger reading did not refer in this paper? ____
   A "Danger"　　　　　　　　　　　　B. "Electric Shock"
   C. "High Voltage"　　　　　　　　　D. "Keep Off"

II. Answer the following questions according to the next.

1. What method did contractors use to determine lengths?
2. Why does the damper may need to be placed in a subspan?
3. What are embedded pads used for?

III. Translate the following into Chinese.

Jumper cables are used to connect conductors across dead-end towers. The calculation of jumper cable lengths varies, with each utility developing its own method empirically. Theoretical models are difficult because jumper cables are too small to approach a catenary or parabolic shape. Therefore computer models or graphical methods are employed to determine jumper lengths. A field method used by contractors is to climb the tower, simulate the jumper by using a rope, gage the correct length by measuring clearances achieved, then cut the conductor to match the rope length. Not very sophisticated, but it produces acceptable results.

Keys

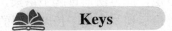

I. Choose the best answer into the blank.

1. D　　2. A　　3. B

II. Answer the following questions according to the next.

1. A field method used by contractors is to climb the tower, simulate the jumper by using a rope, gage the correct length by measuring clearances achieved, then cut the conductor to match the rope length.

2. The damper may need to be placed in a subspan because any device attached to the conductor that weighs more than a length of 6ft (1.8m) of the conductor will reflect vibrations.

3. Embedded pads are used for soils that cannot be augered (such as areas heavy in boulders or in areas of collapsing sand).

III. Translate the following into Chinese.

跳线用于连接穿过终端尾杆塔的导线，其长度计算方法有多种，这些方法都是凭经验得到的。由于理论模型中跳线太小，接近一个悬链或抛物线形状，运用理论模型计算跳线长度较为困难，因此，通常采用计算机模型或图解方法来得到结果。建筑承包商常采用一种现场方法，此方法通过爬上塔架，用绳子模拟跳线，再通过测量绳子来获取正确的跳线长度，最后切割与绳子长度相同的导线，即为所求跳线。该方法过程较为简单，结果也满足工程要求。

翻译技巧之否定句（一）

在英语的否定结构中，由于习惯用法问题，其中部分否定句所表示的意思是不能按字面顺序译成汉语的，因此，翻译时要特别注意。

1. 部分否定

英语中含有全体意义的代词和副词如 all，every，both, always，altogether，entirely 等统称为总括词。它们用于否定结构时不是表示全部否定，而只表示其中的一部分被否定。因此，汉译时不能译作"一切……都不"，而应译为"并非一切……都是的"，或"一切……不都是"。例如：

All of the heat supplied to the engine is not converted into useful work. 并非供给热机的所有热量都被转变为有用功的。错译：所有供给热机的热量都没有被转变为有用功的。

Every one cannot do these tests. 并非人人都能做这些试验。错译：每个人都不能做这些试验。

Both instruments are not precise. 两台仪器并不都是精密的。错译：两台仪器都不是精密的。

This plant does not always make such machine tools. 这个工厂并不总是制造这样的机床。错译：这个工厂总是不制造这样的机床。

2. 全部否定

但是当（这些总括词 + 肯定式谓语 + 含否定意义的单词……）时，则是表示全部否用。not 否定谓语动词是常见的一种全部否定形式。在翻译时一般都译成否定谓语，这与汉语的否定结构基本相同，除了 not 以外，其他表示全部否定意义的词有：no、nobody、none、nowhere、never、neither、nor、nothing，不管这些表示否定意义的词在句中作主语、宾语还是其他成分，这类句子通常译成否定句。例如：

All germs are invisible to the naked eye. 一切细菌都是肉眼看不见的。

Both data are incomplete. 两个数据都不完整。

Nothing in the world moves faster than light. 世界上没有任何东西比光传得更快了。

Section 2　Externality Identification and Quantification of Transmission Construction Projects

1. Introduction

Externality is those effects which are not reflected directly in the production and consumption. Externality exists in the computer, railway, telecom and power industry. It has been seldom to consider the externality during the evaluation job so far, however there is evidence that the externality may have a great influence on the decision of the projects. That is to say, even some projects are economically feasible according to the economic analysis, they may have negative externalities which are harmful to the whole society and people outside the project, as a result it calls for reconsidering whether these projects should be operated. Moreover, people become more and more concerned about the environment and market efficiency. So it is important and urgent to find ways to identify and quantify the externalities of the projects to be constructed.

Like other network industry, power industry has some characteristics such as natural monopoly, public good property and externality. However, due to its importance in the social production and daily life, and its special market environment, power industry has some specific characteristics. As a result, it requires consideration of the special characteristics and the deregulated power market environment when studying the externality of the construction projects in the power industry.

2. Externality of the Transmission Construction Projects

(1) Features of externality.

Positive and negative externalities are the two types of the externalities. It is positive when the external parties benefit from someone's activity and it is negative when external parties lose. Metcalfe Rule named by the inventor of Ethernet Metcalfe can be used to illustrate the network externality vividly. It points out that the value of the network is directly proportional to the amounts of the consumers. The economic meaning of the rule is network externality. Network externality implies the increasing value of some behavior and this value increasing phenomenon goes with the increase of the amount of the participants.

There are some vivid examples of the network externalities. If more people use the E-mail through the telecommunication network, the value of E-mail will increase, that is because the former users can get external profit due to the increase of the potential communicators. Congestion is an example of the negative externality. It will congest if the amount of the E-mail users rises to some extent. The users will suffer from the low speed of the telecommunication network which is the negative externality.

(2) Externality of the transmission construction projects.

Transmission network is the core of the power system and through which the power flow transmits. Connecting the generators with consumers, the transmission network makes the power system integrated. Transmission network also has externalities.

The transmission construction projects have some positive externalities such as the interconnection of the neighboring grids can reduce the reserve capacity and the generation construction investment. Thanks to the transmission construction projects, the transmission system can conduct peak shaving and reserve sharing for unexpected events, and dispatch the hydropower in the neighboring power system so as to enhance the security and reliability of the power systems. However, the transmission construction projects produce negative externalities.

Due to the increase of the consumers and the market power execution behaviors of some participants, the existing transmission capacity no longer meets the demand of the consumers and causes transmission congestions. Moreover, if outage happens on one transmission line, it will affect the power supply of other lines or the whole power system. The transmission construction project will produce noise, take up land which are farms or communities, consume the resource, be harmful to the environment. The electromagnetic radiation of the transmission equipments may damage people's health. All these are negative externalities of the transmission construction projects.

3. Identify the Influence on the External Environment

(1) Impact on the land: It takes large quantity of land to construct the transmission equipments. The land may include farms, forest or dwelling houses. This paper uses the cost pay for the habitants to quantify the impact of the construction projects on the land. In China, the cost is mainly comprised of three parts: compensation fees for taking up the land, subsidies for rehousing and compensation fees for housebreaking. In accordance with the Land Management Law of People's Republic of China, the compensation fees for taking up the land are 6 to 10 times of the average output of the former three years before it is taken up. Subsidies for rehousing are calculated according to the agricultural population, and they are 4 to 6 times of the average output of the three years before it is taken up. The compensation fees for housebreaking are appropriated according to the building's market value.

(2) Noise: Among all the grid construction projects, taking the construction of substation as an example, the transformer, reactor, axial fans of a substation will make noise during their operation and some kind of the noises are hard to eliminate. All the equipments should comply with the related regulation about noise. This paper uses the average degree of the noise in the construction area to quantify the noisy level of the transmission construction projects.

(3) Impact on the vegetation and soil: The transmission construction projects will dig and press the land which will destroy the vegetation and soil. We can use the acreage of damaged vegetation and change rate of the green land per person to calculate the impact of transmission construction projects on the soil.

(4) Impact on the wild life: The construction and operation of the transmission projects will affect the habitat areas of the wild life and even endanger their lives. It is an efficient way to use the reduced amount of the wild life and the percentage of the extinct species to illustrate the impact of transmission construction projects on the wild life.

(5) Impact on human health: The electromagnetic wave emitted by the transmission equipments may harm the human health and cause some chronic diseases. The workers who install the equipments expose to the danger of getting hurt or losing life. The mortality and morbidity of the habitants and workers can be used to approximately evaluate the impact of the transmission construction projects on human health.

原文翻译

第二节 输电线路建设项目的外部性识别和量化

1. 引言

外部性是指不能直接反映生产和消费的影响因素，它存在于计算机、铁路、电信和电力行业中。然而，至今在工作过程中很少考虑外部性，但有证据表明，外部性可能对项目的决策有很大的影响。换言之，有些项目据分析存在经济可能性，但他们可能具有负外部性，进而有害于整个社会和人民，最终导致该项目的可行性需要重新考虑。此外，人们越来越关注环境和市场效率，因此，寻求识别和量化项目外部性的方法是非常重要和迫切的。

类似于其他的网络行业，电力行业具有一定的特性，如自然垄断、公益属性和外部性。然而，由于其在社会生产和日常生活中的重要性，以及特殊的市场环境，电力行业具有一些特殊性。因此，在研究电力行业建设项目的外部性时，需要考虑它的特殊性和放松管制的电力市场环境。

2. 输电线路建设项目的外部性

（1）外部性特点。

正、负外部性是两类外部性。当别人的活动使外部各方受益时即为正外部性，使外部各方受损时即为负外部性。由以太网的发明者梅特卡夫命名的梅特卡夫法则可以非常生动的说明网络外部性。它指出，网络的价值和消费者的数量成正比。这个法则的经济意义正是网络外部性。网络外部性意味着增加一些行为的价值，这个价值将随着参与者数量的增加而增加。

这里有一个关于网络外部性的生动事例。如果越来越多的人通过电信网络使用电子邮件，由于潜在传播者的增加使得最开始的用户能获得额外益处，因此电子邮件的价值就会增加。信息交通拥堵是其负外部性的一个例子。如果电子邮件用户量上升到某种程度上，电信网络将会拥塞。此时，用户需承受电信网络速度的降低带来的影响，这就是负面的外部性。

（2）输电线路建设项目的外部性。

输电网起传输电能的作用，是电力系统的核心，其连接着发电机与消费者，使得电源系统集成化。它同样也具有外部性。

输电建设项目有一定的正外部性，例如周边电网的互联可减少备用容量和发电建设投资。归功于输电的建设项目，输电系统可以对突发事件进行调峰和储备共享，并调度相邻电力系统的水电，以便提高电力系统的安全性和可靠性。然而，输电建设项目也存在负外部性。由于消费者数量和一些参与者市场执行行为的增加，使得现有的传输容量不再满足消费者的需求，因而会导致输电拥塞。此外，如果一条输电线发生断线，就会影响其他线路或整个电力系统的电能传输。输电建设项目还会产生噪声，占用农场或社区土地，消耗资源，对环境有害。同时，输电设备的电磁辐射也可能会损害人体健康。所有这些都是输电建设项目的负外部性。

3. 确定对外部环境的影响

（1）对土地的影响：建设输电设备需要大量的土地。这些土地可能包括农场、林场或住宅。本文采用居民成本来量化建设项目对土地的影响。在中国，这个成本主要由三部分组成：占用土地补偿费、重新安置补贴及拆房补偿费。按照中华人民共和国土地管理法，土地补偿费按占用该地块前三年年平均产值的 6 至 10 倍发放。重新安置补贴按农业人口计算,按占用该地块前三年平均年产值的 4 至 6 倍发放。拆房补偿费是按照建筑的市场价值发放。

(2)噪声的影响:在所有的电网建设项目中,以建设变电站为例,变电站中变压器、电抗器、轴向电扇在运行过程中都会产生噪声,而这些噪声中有些难以消除。所有设备应符合有关噪声的有关规定。本文使用建筑区域的平均噪声程度来量化传输建设项目的噪声水平。

(3)对植被和土壤影响:输电建设项目对土地的挖掘和重压会破坏植被和土壤。我们可以用被破坏植被的面积和变化人均绿地率来计算输电建设项目对土壤的影响。

(4)对野生动物的影响:输电项目的建设和运营将影响野生动物的栖息地,甚至危及其生命。野生动物减少的数量和灭绝物种的比例可以有效说明输电建设项目对野生动物的影响。

(5)对人体健康的影响:输电设备辐射的电磁波可能会损害人的健康,并导致一些慢性疾病的产生。安装设备的工人随时都面临着受伤,甚至是失去生命的危险。死亡率和居民与工人的发病率可以用来近似评估输电建设项目对人体健康的影响。

New Words and Expressions

hydropower	*n.* 水力发出的电
power industry	电力工业
power flow	电流
electromagnetic radiation	电磁辐射
power market	电力市场

Notes

1. As a result, it requires consideration of the special characteristics and the deregulated power market environment when studying the externality of the construction projects in the power industry.

因此,在研究电力行业建设项目的外部性时,需要考虑它的特殊性和放松管制的电力市场环境。

2. Transmission network is the core of the power system and through which the power flow transmits. Connecting the generators with consumers, the transmission network makes the power system integrated. Transmission network also has externalities.

输电网起传输电能的作用,是电力系统的核心。输电网络连接着发电机与消费者,使得电源系统集成化。它同样也具有外部性。

3. The transmission construction projects have some positive externalities such as the interconnection of the neighboring grids can reduce the reserve capacity and the generation construction investment.

输电建设项目有一定的正外部性,例如周边电网的互联可减少备用容量和发电建设投资。

4. Among all the grid construction projects, taking the construction of substation as an example, the transformer, reactor, axial fans of a substation will make noise during their

operation and some kind of the noises are hard to eliminate.

在所有的电网建设项目中，以建设变电站为例，变电站中变压器、电抗器、轴向电扇在运行过程中都会产生噪声，而这些噪声中有些难以消除。

Exercises

Ⅰ. Choose the best answer into the black.

1. Externality is those effects which are not reflected ___ in the production and consumption.

  A. directly   B. direct   C. indirect   D. indirectly

3. It is important and urgent to find ways to identify and quantify the ___ of the projects to be constructed.

  A. use   B. mean   C. externalities   D. noise

3. Power industry has some characteristics such as natural monopoly, public ___ property and externality.

  A. good   B. bad   C. neutral   D. objective

Ⅱ. Answer the following questions according to the text.

1. Power industry has some characteristics.give some example.

2. What are negative externalities of the transmission construction projects?

3. Talk about the impact on the wild life with the construction and operation of the transmission projects.

Ⅲ. Translate the following into Chinese.

The electromagnetic wave emitted by the transmission equipments may harm the human health and cause some chronic diseases. The workers who install the equipments expose to the danger of getting hurt or losing life. The mortality and morbidity of the habitants and workers can be used to approximately evaluate the impact of the transmission construction projects on human health.

Keys

Ⅰ. Choose the best answer into the black.

1. A   2. C   3. A

Ⅱ. Answer the following questions according to the text.

1. Natural monopoly, public good property and externality.

2. transmission congestions,affect the power supply of other lines or the whole power system,produce noise, take up land which are farms or communities, consume the resource, be harmful to the environment.

3. The construction and operation of the transmission projects will affect the habitat areas of the wild life and even endanger their lives.

Ⅲ. Translate the following into Chinese.
输电设备辐射的电磁波可能会损害人的健康，并导致一些慢性疾病的产生。安装设备的工人随时都面临着受伤，甚至是失去生命的危险。死亡率和居民与工人的发病率可以用来近似评估输电建设项目对人体健康的影响。

翻译技巧之否定句（二）

1. 双重否定

双重否定结构通常是由 no, not , never, nothing 等词与含有否定意义的词连用而构成的。这种结构形式上是否定，实质上是肯定，语气较强。 翻译时可译为双重否定，有时也可译成肯定句。

You can do nothing without energy.没有能量，你就什么也做不成。

In fact , there is hardly any sphere of life where electricity may not find useful application.事实上，几乎任何一个生活领域都要用到电。

2. 意义上的否定

英语中有些词和词组在意义上表示否定。如：little（几乎没有），few（几乎没有），seldom（极少），scarcely（几乎不），hardly（很难，几乎不），too……to（太……以致不……），rather than（而不）fail to（不成功……未能……）等。翻译时要译出否定的意义。

Metals, generally, offer little resistance and are good conductors.通常金属几乎没有电阻，因而是良导体。

Glass conducts so little current that it is hardly measurable.玻璃几乎不导电，因此很难测量其中的电流。

3. 否定转移

因英语与汉语在表达否定要领时所使用的词汇手段与语法手段都有很大的差别，所以在翻译时常用到两种转移：一是语法的转移，即否定主语或宾语转移成否定谓语，否定谓语转移成否定状语等。二是内容上的否定转移，即英语中的否定形式译成汉语时可用肯定形式，反之亦然。

No smaller quantity of electricity than the electron has ever been discovered. 从来没有发现过比电子电荷更小的电量。（由否定主语转移为否定谓语）

Electric current cannot flow easily in some substances.电流不能顺利地在某些物质中流动。（从逻辑上判断是否定状语，easily）

Section 3 Some Introductions about The Transmission Line Construction

1 Installation of ground wire, electrodes and conductors

Twenty-eight days after the footings are installed, the ground wire, electrodes and conductors may be strung and sagged into place by the construction contractor as supervised by the quality assurance engineer.

Ground wire and conductor stringing and sagging are usually carried out between successive dead-end towers. This is because dead-end towers are designed to withstand unbalanced loadings that occur during these operations. However, where a long run of successive suspension towers occurs with no intervening dead-end tower, suspension towers may be backguyed and stringing and sagging operations carried out between suspension towers. Vehicle mounted pulling and tensioning equipment is stationed at each end of the pull. Stringing hardware is installed on the towers prior to the operation, and sagging personnel are placed in every fourth or fifth tower, as needed. A polypropylene line is strung through the stringing hardware and connected to a 5/8 in. (1.59cm) steel pulling cable which is pulled through the stringing hardware. The conductor is connected to the pulling cable and pulled through. The conductor hardware is then connected to each dead-end tower. The conductor is removed from the stringing hardware, sagged, and connected to the suspension towers once clipping offset results are implemented. Sags are measured by setting sights on the structures at each end of the span at a vertical distance below the conductor support equal to the sag.

When the transmission line crosses any structure or other line, the following practices should be adhered to.

(1) Permits from the local regulating agency are required to cross any thoroughfare.

(2) When crossing any regularly traveled thoroughfare, guard structures must be erected prior to any stringing operations. The guard structure consists of two sets of wood pole structures. Each wood pole structure consists of two poles with a crossarm connecting them.

(3) When crossing a major thoroughfare, a net must be strung, in addition to the guard structures described above.

(4) When crossing a communication or power line, crossing permits must be obtained from the existing line's utility. A temporary outage during crossing is preferred.

(5) Some lines, for example most 500kV lines, cannot tolerate even a temporary outage for crossing operations. When crossing an energized line, extra conditions must be worked out with the existing line's utility. In some cases, cranes with travelers are used to facilitate crossings.

2 Vibration dampers

Wind, or vibration, dampers are required to prevent damage from strain fatigue to the conductors from aeolian vibration. This vibration, which ranges from 5 to 150 cycles per second, is caused by transverse winds creating eddies on the leeward side of the conductor which swing from the upper to the lower side of the conductor at regular intervals. The vibration rate depends on the diameter of the conductors. For this reason, large-diameter lightweight conductors are especially prone to vibration. ACSR conductor is relatively light and is usually strung to high tensions and thus is quite susceptible to vibration.

Therefore this kind of conductor requires special protection by the use of vibratioVibration dampers operate by converting the mechanical energy of conductor motion into dissipative heat energy.The span length and working stress are major factors in wind caused damage. Other factors include attached hardware, conductor diameter, and support structure composition. Wind speeds requiring a damping range from 2 to 30 miles per hour (3～48km/h) with speeds of

2-6 miles per hour (3-10km/h) being of most concern. The maximum safe limit for bending strain in aluminum or ACSR conductors is 150 micro-inch/inch. The maximum safe limit in all-steel conductors (i.e. ground wire) is 225 micro-inch/inch. Dampers are used to keep the bending strain below these limits.

2.1 Installation of vibration dampers

Vibration dampers are installed in spans or subspans where fatigue due to aeolian vibration is predicted. The damper may need to be placed in a subspan because any device attached to the conductor that weighs more than a length of 6 ft (1.8m) of the conductor will reflect vibrations. Thus a damper on the non-reflected side may not be able to adequately damp vibrations.

The placement of the damper is a trade off because placing a damper closer to a reflecting object (i.e. a tower) absorbs more energy, while placing it farther from a tower protects more of the span. In placing dampers, node points are to be avoided. Node points are locations where damping has no effect. There are at least two node points per span. Typical placement of dampers, determined empirically, is given below. These recommendations apply only to Stockbridge type dampers.

2.1.1 Suspension towers

Node locations are known; they are located at points of suspension.

Span length less than 1400ft (427m): use one set of dampers per conductor, placed 5ft (1.5m) from the suspension point.

Span length more than 1400ft (427m): use two sets of dampers per conductor. Place one set 5ft (1.5m) from the suspension point and the second set 7ft (2.1m) further along the span.

2.1.2 Dead-end towers

Node locations are not known.

Span length less than 1400ft (427m): use two sets of dampers per conductor. Place both sets by one dead end, one set 5ft (1.5m) from the dead-end compression assembly and the second set 7ft (2.1m) further along the span.

Span length more than 1400ft (427m): use four sets of dampers per conductor. Place two by each dead end. At each dead end place one set 5ft (1.5m) from the dead-end compression assembly and the second set 7ft (2.1m) further along the span.

原文翻译

第三节　输电线路施工简介

1　地线、电极和导线的安装

杆塔基础安装完毕 28 天后,将接地线、接地极和导线由施工承包商串接和下垂到位,全过程由质量保证工程师监督。

接地线和导线的串接、下垂通常在连续的尾杆塔之间进行,这是因为在操作过程中尾杆塔能够承受不平衡荷载。然而，在没有尾杆塔干扰的区域，长期运行的连续耐张塔可能会被导线拉倒，故耐张塔之间需进行串接和下垂操作。利用汽车分别将牵引和张紧设备安装在拉线两端。在操作之前，需将绞线硬件安装在杆塔上，并且根据需要每隔四基或五基杆塔放置一个下垂操作人员中。利用串接硬件对聚丙烯线路进行串接，再将线路通过串接硬件连接到牵引钢绳 5/8 英寸（1.59cm）处并向前拉。随后，将导线连接器具连接到各尾杆塔上。剪切补偿结果一旦实施，将导线从放线器具上移走，然后进行下垂操作，从而连接到耐张塔上。在挡距末端结构上设置观测点，该观测点位于绝缘子下端，其于弧垂最低点平行，弧垂即可以通过观测点得到。

当输电线路穿过任何建筑物或其他线路时，必须遵循以下惯例。

（1）当线路需穿越要道时，必须有当地监管机构的许可。

（2）当穿过任何规划通行的大街时，必须在所有串线操作之前架设防护装置。该装置由两套木杆组成，而每套木杆由两个连接有横木的杆组成。

（3）当通过主干道时，除了上面描述的保护装置外，还必须架设网状保护系统。

（4）当跨接通信线路或其他线路时，必须获得现有线路效用过境许可证。并且在此时最好采取临时断电措施。

（5）有些线路，例如多数 500kV 线路不能容许在跨接操作中出现暂时断电，当通过带电线路时，必须用现有的线路功能计算出额外情况。在某些情况下，需要利用起重机来辅助完成跨接操作。

2　减振器

有风或存在振动时，必须采用阻尼器来防止由微风、振动引起的导线疲劳损坏。横向风会在导线背风侧产生漩涡，从而使导线每隔一定时间都将产生从上到下振动，这种振动每秒产生 5 到 150 次。该振动频率取决于导线的直径。因此，大直径轻型的导线尤其容易振动。钢芯铝绞线相对来说比较轻，并且经常用来串接高电压，因而更容易产生振动。

因此这种导线需要采用振动模式的振动阻尼器来进行特殊保护，这类减振器主要是通过运动将导线机械能转换为热能来工作的。风中的挡距和工作应力是造成导线损坏的主要因素。其他因素包括附加的金具、导线的直径和支撑结构的成分。需要减震的风速范围为每小时 2～30 英里（3～48km/h），其中最值得关注的是每小时 2～6 英里（3～10km/h）的风速。铝线或钢芯铝绞线的最大安全弯曲应力限制在 150 微英寸/英寸以内。所有钢绞线（如接地线）的最大安全弯曲应力限制在 225 微英寸/英寸以内。阻尼器就是用来保持弯曲应力在这些限度以下的。

2.1　减振器的安装

减振器通常安装在预测风振可能引起疲劳的挡距或次挡距中。由于任何超过 6 英尺（1.8m）导线重量的导线附加设备都将产生振动，因此，减振器可能需要放置在次挡距上。需要注意的是，在没有振动一侧的减振器可能无法充分的进行衰减振动。

减振器的安置是一个折中方案，因为把减振器安置在靠近易振动的物体（如塔）时会吸收更多能量，而放在离塔较远的地方却能够保护更多的挡距。由于节点是减振器不起作用的位置，因此减振器不能安装在节点处。每个挡距上至少有两个节点。根据经验，减振器安装的一般位置如下。值得注意的是，这些位置只适用于斯托克型减振器。

2.1.1　耐张塔

节点位置已知；它们位于悬挂点处。

挡距小于 1400ft（427m）：每根导线用一套减振器，安装在离悬挂点 5ft（1.5m）处。

挡距超过 1400ft（427m）：每根导线用两套减振器，一套安装离悬挂点 5ft（1.5m）处，另一套安装离挡距 7ft（2.1m）处。

2.1.2 终端塔

节点位置未知。

挡距小于 1400ft（427m）：每根导线用两个套减振器，把两套安装于同一挡端处。其中，一套安装在离挡端 5ft（1.5m）处，另外一套安装离挡距 7ft（2.1m）处。

挡距超过 1400ft（427m）：每根导线用四套减振器，将两套分别放在两挡端处。其中，两套分别安装在离两挡端 5ft（1.5m）处，而另两套则分别安装在离两挡端 7ft（2.1m）处。

  New Words and Expressions

cable	*n.* 电缆，电报
crossarm	*n.* 电线杆上的横木
dead-end	终端

 Notes

1. Twenty-eight days after the footings are installed, the ground wire, electrodes and conductors may be strung and sagged into place by the construction contractor as supervised by the quality assurance engineer.

杆塔基础安装完毕 28 天后，将接地线、接地极和导线由施工承包商串接和下垂到位，全过程由质量保证工程师监督。

2. Vibration dampers are installed in spans or subspans where fatigue due to aeolian vibration is predicted.

减振器通常安装在预测风振可能引起疲劳的挡距或次挡距中。

3. The placement of the damper is a trade off because placing a damper closer to a reflecting object (i.e. a tower) absorbs more energy, while placing it farther from a tower protects more of the span. In placing dampers, node points are to be avoided.

减振器的安置是一个折中方案，因为把减振器安置在靠近易振动的物体（如塔）时会吸收更多能量，而放在离塔较远的地方却能够保护更多的挡距。由于节点是减振器不起作用的位置，因此减振器不能安装在节点处。

 Exercises

Ⅰ. Choose the best answer into the black.

1. Ground wire and conductor stringing and sagging are usually carried ____ between successive dead-end towers.

  A. of   B. out   C. in   D. with

2. Wind, or vibration, dampers are required to prevent damage from strain fatigue to the conductors from ___ vibration.

    A. aeolian   B. sand   C. cold   D. filthy

3. The ___ safe limit for bending strain in aluminum or ACSR conductors is 150 micro-inch/inch.

    A. minimum   B. maximum   C. medium   D. high

Ⅱ. Answer the following questions according to the text.

1. What are major factors in wind caused damage?

2. What are installed in spans or subspans?

Ⅲ. Translate the following into Chinese.

Vehicle mounted pulling and tensioning equipment is stationed at each end of the pull. Stringing hardware is installed on the towers prior to the operation, and sagging personnel are placed in every fourth or fifth tower, as needed. A polypropylene line is strung through the stringing hardware and connected to a 5/8 in. (1.59cm) steel pulling cable which is pulled through the stringing hardware. The conductor is connected to the pulling cable and pulled through. The conductor hardware is then connected to each dead-end tower. The conductor is removed from the stringing hardware, sagged, and connected to the suspension towers once clipping offset results are implemented. Sags are measured by setting sights on the structures at each end of the span at a vertical distance below the conductor support equal to the sag.

Keys

Ⅰ. Choose the best answer into the black.

1. B   2. A   3. B

Ⅱ. Answer the following questions according to the text.

1. The span length and working stress.

2. Vibration dampers.

Ⅲ. Translate the following into Chinese.

利用汽车分别将牵引和张紧设备安装在拉线两端。在操作之前，需将绞线硬件安装在杆塔上，并且根据需要每隔四基或五基杆塔放置一个下垂操作人员。利用串接硬件对聚丙烯线路进行串接，再将线路通过串接硬件连接到牵引钢绳 5/8（1.59cm）处并向前拉。随后，将导线连接器具连接到各终端塔上。剪切补偿结果一旦实施，将导线从放线器具上移走，然后进行下垂操作，从而连接到耐张塔上。在挡距末端结构上设置观测点，该观测点位于绝缘子下端，其于弧垂最低点平行，弧垂即可以通过观测点得到。

翻译技巧之定语从句

（1）只要是比较短的，或者虽然较长，但汉译后放在被修饰语之前仍然很通顺，一般的就放在被修饰语之前，这种译法叫作逆序合译法。例如：

The speed of wave is the distance it advances per unit time. 波速是波在单位时间内前进的距离。

The light wave that has bounced off the reflecting surface is called the reflected ray. 从反射表面跳回的光波称为反射线。

Stainless steel, hick is very popular for its resistance to rusting, contains large percentage of chromium. 具有突出防锈性能的不锈钢含铬的百分比很高。

（2）定语从句较长，或者虽然不长，但汉译时放在被修饰语之前实在不通顺的就后置，作为词组或分句。这种译法叫作顺序分译法。例如：

Each kind of atom seems to have a definite number of "hands" that it can use to hold on to others. 每一种原子似乎都有一定数目的"手"，用来抓牢其他原子。

（顺序分译法）每一种原子似乎都有一定数目用于抓牢其他原子的手。

（逆序合译法）这句限制性定语从句虽然不长，但用顺序分译法译出的译文要比用逆序合译法更为通顺。

Let AB in the figure above represent an inclined plane the surface of which is smooth and unbending.

设上图中 AB 代表一个倾斜平面，其表面光滑不弯。（顺序分译法）

设上图中 AB 代表一个其表面为光滑不弯的倾斜平面。（逆序合译法）

上面两种译法，看来也是用顺序分译法比用逆序合译法更为通顺 简明。

（3）定语从句较长，与主句关联又不紧密，汉译时就作为独立句放在主句之后。这种译法仍然是顺序分译法。例如：

Such a slow compression carries the gas through a series of states, each of which is very nearly an equilibrium state and it is called a quasi-static or a " nearly static" process. 这样的缓慢压缩能使这种气体经历一系列的状态，但各状态都很接近于平衡状态，所以叫作准静态过程，或"近似稳定"过程。

Friction wears away metal in the moving parts, which shortens their working life. 运动部件间的摩擦力使金属磨损，这就缩短了运动部件的使用寿命。

（4）There + be 句型中的限制性定语从句汉译时往往可以把主句中的主语和定语从句融合一起，译成一个独立的句子。这种译法叫作融合法，也叫拆译法。例如：

There are bacteria that help plants grow, others that get rid of dead animals and plants by making them decay, and some that live in soil and make it better for growing crops. 有些细菌能帮助植物生长，另一些细菌则通过腐蚀来消除死去的动物和植物，还有一些细菌则生活在土壤里，使土壤变得对种植庄稼更有好处。

There is a one-seated which you could learn to drive in fifty minutes. 有一种单座式汽车，五十分钟就能让你学会驾驶。

Chapter 6　Typical Defects from Power Transmission Line

Section 1 An Overview of the Condition Monitoring of Overhead Lines

The bulk of the electricity consumed in England, Wales and Scotland is transported by the 275 and 400kV network of overhead lines. Approximately half the routes have twin bundled conductors and the other half, generally the newest have quad bundles. A major part of the 7000km route involved was built between 1955 and 1970. Some circuits are therefore reaching middle age, i.e. they are between 25 and 40 years old. Individual components such as insulators, joints and suspension clamps have deteriorated and they have to be repaired or replaced as part of normal maintenance. Short sections of a conductor can also be repaired or replaced when damaged.

In the drive for greater efficiency, particularly in terms of reduced outage times, power utilities and equipment manufacturers are actively seeking new ways to reduce the cost of design, development and maintenance of equipment without adversely affecting the reliability and safety of the transmission system. But degradation as a result of ageing under service conditions is inevitable. The problem of how to identify and quantify this deterioration so that a reliable system can be maintained is made difficult by the problems of physical access to the line components. There is thus a requirement to seek new methods to enable prompt identification of defective equipment under service conditions, i.e. through on-line processing for monitoring the health of equipment, commonly known as condition monitoring (CM).

This paper reviews the main deterioration problems that have been found in overhead lines including the fine cracks on insulators, corrosion of the conductor and mechanical damage to conductors by wind-induced movement, and analyses the causes of these problems. It then discusses the various detection techniques which are presently being used or may be implemented to improve transmission line inspections. Since most of these techniques require mobilisation by foot, ground or air vehicles, none of them can be realised as on-line condition monitoring. The paper finally outlines a current collaborative research project involving the University of Bath and two major utilities in the UK in the on-line condition monitoring of EHV transmission lines. The aim of this research project is to perform a series of scientifically rigorous tests and theoretical analyses aimed at defining and elucidating the effects which the introduction of defective conductors or cracked insulators have on the partial discharge characteristics of overhead lines and thereby assist in the design and engineering of an on-line condition monitoring device having a performance that is superior to other techniques through the use of artificial intelligence (AI) techniques.

1. Problems of deterioration in transmission lines

Problems of deterioration in transmission lines over which electrical power is transmitted have been continuously increasing along with system voltage levels. This has led to a significant increase in the number of suspension insulators in use on power lines. Insulation is largely responsible for the operating performance of a transmission line. Safe design requires a dry flashover of the insulator string of three to five times the nominal operating voltage and a leakage path approximately twice the strike distance. These design practices are modified for special cases such as salt, fog, dust, or chemical laden air.

Many countries use strings of cap and pin insulators, sometimes called discs, for supporting the conductors of overhead power lines. The insulator shell, of either glass or porcelain, is fixed by cement to a metal cap and a metal pin. The shed is shaped and sized so as to achieve a high electric strength across its surface. The defective rates are much higher for the 400kV lines (normally 22 unit suspension string) than for the 275kV lines (usually 16 unit suspension string); for generally similar locations, the highest failure rates are for lines near to the sea. Most defective units are at the high voltage end of the string; in the worst case, one third of the conductor end units are defective.

Transmission line conductors are the most important and the most expensive components of EHV lines; they are, however, susceptible to ageing. One of the main reasons limiting their lifetime is unavoidable wind induced aeolian vibrations. Many important lines in several countries are now reaching an age where the mechanical safety tends to unacceptably low values, a condition which is very often little understood. Also, due to the increasing use of new types of conductors (e.g. new ACSR stranding, new materials, AAAC, ACAR, Alumoweld) where long term design experience does not exist, serious conductor fatigue and failures occur, causing increasing concern.

Some line condition assessment work has already been carried out on the British Transmission System.The aim is to assess the condition of overhead lines by non-invasive means. The main problems that have been considered are the cracks in porcelain insulators and broken conductor strands. Similar problems have also been found in overhead lines in other countries.

2. Damage to insulators

The extent of the cracked insulator problem has been increasingly realised since the 1970s, with the recognition that mechanical factors of safety have declined. An insulator comes to the end of its working life either when it fails mechanically, flashes over at unacceptably high frequencies or gives evidence of deterioration to a condition likely to lower its factor of safety in service. All insulators are affected to some extent by impact, thermal and mechanical cycling, ablation from weathering and electro-thermal causes, flexure and torsion, ionic motion, corrosion and cement growth. Looms has summarised that the cap and pin disc porcelain insulators are mainly damaged by cycling, cement growth and corrosion.

Cyclic loading is a known cause of material failure since it promotes both growth of micro-cracks directly and allows ingress of water into all kinds of surface flaw. In cap and pin disc insulators, care is taken to accommodate cyclic changes by mechanical design. In the case of thermal cycling, the relative expansibility of tile metal fittings, cement and dielectric determines the sizes of stress which are generated, not only by temperature excursions from full sunshine to clear night sky, which may exceed 80℃, but heat generation under passage of fault current arcs.

3. Identification of problem symptoms

The symptoms associated with each problem on overhead lines have been identified and quantified in an EPRI report. Although a single problem may display more than one symptom, the generation of partial discharge (PD) is considered as a major and sensitive factor employed in many of the problems that are being investigated. PDs generally occur whenever a transmission line is energised. Depending upon the weather, age of the line, problem conditions, and other factors, the level of discharge can be predicted as well as measured. This is more so in view of the fact that many of the problem conditions may raise the level of discharge generation above the expected or normal level of the line.

4. An assessment of the techniques for inspection of lines

A number of papers have dealt with the techniques of overhead line inspections. High frequency partial discharges have been considered as a main symptom with many of the problems that have been investigated, which theoretically generate radio noise into the ultra-high frequency range and audible noise into the ultra-sonic regions. Above normal heat generation is another symptom. Thus all the sensing systems have focused on partial discharge levels and temperatures. The main detection techniques are summarised as follows:

(1) Ultrasonic detection.

(2) Measurement of corona pulse current inconsistency.

(3) Partial discharge detector.

(4) Infrared inspection of overhead transmission lines.

(5) Radio noise detection system.

(6) Solar-blind power line inspection system.

(7) Corona current monitor for H.V. power lines.

(8) Fibre optic applications to transmission line inspections.

(9) Audible noise meters.

(10) Field testing of insulators.

Most of these techniques require mobilisation by foot, ground vehicle, wire tracking vehicle or air vehicle, and involve human presence in the vicinity of the line. Safety considerations and the dependency on the human senses and subjectivity of inspection or live-line maintenance crews, in the noisy and physically demanding environment are serious limitations to take full advantage of all the available technologies. None of these techniques can monitor the entire lines automatically and in real time by the installed detectors at the two ends of a line. The most efficient method of inspecting a transmission line to date has been by aerial inspection, particularly using a helicopter. A helicopter has hovering capability (an advantage), whenever a problem is located and a second look or closer investigation is desired.

5. A novel condition monitoring technique

It is widely accepted that defective conductors, weak or polluted insulators can lead to permanent breakdown and hence an unscheduled outage if left undetected. Unplanned outages are often very disruptive operationally, particularly where the transmission lines are connected to nuclear power plant; this is so because in this case, the company has a contractual obligation to guarantee maintenance of line connection. Therefore, reliable and accurate techniques for failure detection, location, positive identification and severity evaluation, have become essential to deal with the urgent maintenance work on time and at a reasonable cost. However, at present there is no wholly satisfactory method of CM available, thus the co-operation among the University of Bath and two major UK utilities, has been established to research into such a technique. It is the intention that this project will, inter-alia, help to ameliorate these difficulties if only to the extent of giving early warning of first failure emanating from beneath spacers, often caused by the breaking of a few conductor strands, or the initial cracks in one or two discs of an insulator string; such conditions often elude detection and a means of effecting even a limited improvement in this situation would be extremely beneficial economically. For example, the total removal of (or simply moving to another point on the conductor) a defective spacer, could considerably extend the conductor life; alternatively, repairs to the damaged part of the conductor could prolong its life for another 10 years or so at a relatively low cost.

第一节　架空输电线路状态监控概述

英格兰、威尔士和苏格兰消耗的大部分电能都是通过275kV和400kV架空线路网传输的。大约一半的线路是两分裂导线，另一半新线路则通常采用四分裂导线。7000km 线路中大部分是在1955年和1970年之间建成的，其中部分线路已经建造了25到40年，其寿命已经过半。一些独立部件，例如绝缘子、接头和吊夹已经老化，作为正常维护的一部分，它们必须要进行修理或者更换。导线只要损坏也需要修理或者更换。

为了追求更大效益，特别是减少断电时间，电力公司和设备制造商都在积极的寻求新方法，旨在不影响输电线路的稳定性和安全性下，减少设计成本以及开发和维护的费用。输电线路即使进行了维护，但由老化引起的性能退化仍然是不可避免的。维持电力系统的可靠性就需要识别和量化这种老化，但直接接近线路元件进行检测较为困难，从而导致这种识别和量化存在很大难度。因此，有必要寻求新的方法来快速识别有缺陷的设备，例如通过在线监控来掌握设备的健康状态，这通常被称作状态监控（CM）。

本文回顾了主要的架空线路恶化问题，包括绝缘子的细裂纹、导体的腐蚀和风振引起的机械损伤，并分析了这些问题的成因。接着讨论了一些目前正在使用的技术或可以提高输电线路巡查的检测技术。由于绝大多数技术均需要步行或使用飞行器，因此，没有一项技术可以实现在线监测。最后，本文概述了目前的合作研究项目，该项目涉及进行超高压输电线路在线监测项目的英国巴斯大学和两大事业单位。该项目的目的是进行一系列科学严格的测试和理论分析，最终定义和阐明架空线路中缺陷导线或破裂绝缘子对局部放电的影响，再结合人工智能（AI）技术，从而设计和建造出优于其他技术性能的在线状态监测设备。

1. 输电线路恶化问题

随着系统电压等级的升高，输电线路的恶化问题也随之增加，最终导致输电线路使用的悬式绝缘子也越来越多，使得绝缘水平很大程度上代表了输电线路的运行性能。出于对安全和盐、雾、粉尘、含有化学成分的空气等因素的影响考虑，绝缘子串干燥闪络电压应为运行电压的三到五倍，而漏电路径长度应为击距的两倍。

很多国家采用帽式或针式绝缘子，它们有时又称圆盘，其目的是支撑架空输电线路。绝缘子外壳，无论是玻璃或是陶瓷的，均是用水泥固定到金属帽和金属针上的。绝缘子伞裙有各种形状和大小，以便提高其表面的电气强度。400kV 的线路绝缘子串（通常是 22 片）缺陷率明显高于 275kV 的绝缘子串（通常 16 片），一般海边线路最容易在类似位置出现故障。缺陷的位置多出现于悬式绝缘子串高压端，而最坏的情况是导线末端三分之一的元件均损坏。

输电线路是超高压线路中最重要，也是最昂贵的一部分。但它们易老化，而老化的主要成因是风引起的振动。一些地区的重要线路已经达到使用年限，其机械安全性已经处于非常低的水平，但这种情况常易被人们忽视。虽然人们越来越多的使用新型导体材料（如新型钢芯铝绞线、新材料、铝合金绞线、铝合金芯铝绞线、铝包钢线等）的导线，但是由于缺乏长期的设计经验，这类导线易出现严重疲劳和故障的情况也逐渐被人们关注。

英国输电系统正在开展线路状况评估工作，旨在通过无伤手段评估输电线路，其主要需要考虑的是瓷质绝缘子出现缝隙、导线断线两类问题。这类问题在其他国家的架空线路上出现过。

2. 绝缘子损坏

自 20 世纪 70 年代以来，由于机械因素的识别能力下降，人们开始越来越重视绝缘子裂纹问题。当绝缘子产生机械问题、高频率的发生闪络或者显示出了一个有可能降低其安全系数的症状，其工作生涯也就随之结束。所有绝缘子都会受一些因素的影响，比如冲击，热力循环，风蚀，电热，弯曲和扭转，粒子运动，水泥增生及腐蚀。柯林斯总结出悬盘式陶瓷绝缘子损坏的主要原因有三个：周期性、水泥增生和腐蚀。

周期性载荷是材料产生故障的原因之一，因为其会直接促使各种细小裂纹的增长，致使水能进入到各种表面裂缝里。在帽式或销盘式绝缘子中，通过严密的机械设计来适应这种周期性问题。在热循环的条件下，陶瓷金具、水泥和介质的相对膨胀性决定了应力的大小，这种应力不仅来自于可能超过 80℃昼夜的温差，同时也来自于故障电流产生的热量。

3. 问题症状识别

美国电力研究协会报告已经发现并量化了关于架空输电线路各种问题的相关症状。虽然一个问题可能会导致多种症状，但在研究中认为最主要、最敏感的因素是局部放电（PD）的成因。一般当输电线路具有高能量时，会发生局部放电。放电程度可以通过天气、线路年限、问题状况和一些其他因素进行预测和测量。鉴于以上事实，许多问题的条件都可能致使线路放电等级比预期或正常水平高。

4. 线路检测的技术评估

现有许多研究架空线路巡线技术的文章均认为高频局部放电是众多问题的主要症状。局部放电理论上会产生超高频范围的无线电噪声和超声区域的可听噪声。高于正常水平的发热则是另外一种症状。因此，所有的传感系统都主要关注局部放电水平和温度，其主要的监测技术总结如下：

（1）超声波探测法。

（2）电晕脉冲电流不一致检测。

（3）局部放电检测器。

（4）架空输电线路红外检测。

（5）无线电噪声监测系统。

（6）太阳能无功在线监测系统。

（7）高压电力线路电晕电流监视。

（8）光纤应用于输电线路检查。

（9）可听噪声测量。

（10）绝缘子现场测试。

上述技术多需要依靠徒步、场地车辆、线路追踪车辆或飞行器，并且还要考虑人对附近线路的影响。在嘈杂和物理要求苛刻的环境中，安全考虑、人的感知依赖性以及人对检查或维护带电线路的主观性严重的限制了所有可用技术的优点。因此，没有一项技术可以通过在线路两端安装监视器实现对整条线路自动、实时检测。迄今为止检测一条输电线路最有效的措施是进行空中检测。该方法多采用直升机，当线路故障需要定位或仔细检查时，直升机盘旋能力能展现较为明显的优势。

5. 一种新的状态监测技术

众所周知，有缺陷的导线，薄弱或遭污染的绝缘子可能导致永久性故障，若未及时检测到这些故障，就会导致意外断电。而计划外的断电是极具破坏性的，特别是当线路连接到核电站时，公司有合同来保证线路正常工作。所以可靠准确地对线路故障进行检测、定位、正确识别及严重程度评估的技术，已经成为及时且价格合理的线路维护中必不可少的一部分。然而，目前没有完全令人满意的 CM 方法可用，所以巴斯大学和两大事业单位开展了对这一技术的研究。这个项目的目标是解决上述困难，希望能从间隔装置下发出故障的早期预警，但故障通常由导线内部断裂或绝缘子串第一片和第二片初次断裂造成的，很难进行检测。该种方法虽然只是在这种情况下进行了有限提高，但在经济上却十分可观。比如说，全部切除（或者在导线上简单移到另一个点）有缺陷的间隙，可以大大延长导线的寿命，或者采用相对较低的成本修理损坏的导线使其多使用 10 年。

New Words and Expressions

degradation	*n.* 退化，恶化
monitoring	*v.* 监控；*n.* 监视器
failure	*n.* 故障，失灵
discharge	*v.* 放电
meters	*n.* 测量仪表
leakage path	漏电路径

short sections   短接
twin bundled conductors   双股线
ACSR   钢芯铝绞线
corona pulse   电晕脉冲

 Notes

1. The problem of how to identify and quantify this deterioration so that a reliable system can be maintained is made difficult by the problems of physical access to the line components.

维持电力系统的可靠性就需要识别和量化这种老化，但直接接近线路元件进行检测较为困难，从而导致这种识别和量化存在很大难度。

2. There is thus a requirement to seek new methods to enable prompt identification of defective equipment under service conditions, i.e. through on-line processing for monitoring the health of equipment, commonly known as condition monitoring (CM).

因此，有必要寻求新的方法来快速识别有缺陷的设备，例如通过在线监控来掌握设备的健康状态，这通常被称作状态监控（CM）。

3. An insulator comes to the end of its working life either when it fails mechanically, flashes over at unacceptably high frequencies or gives evidence of deterioration to a condition likely to lower its factor of safety in service.

当绝缘子产生机械问题、高频率的发生闪络或者显示出了一个有可能降低其安全系数的症状，其工作生涯也就随之结束。

4. Although a single problem may display more than one symptom, the generation of partial discharge (PD) is considered as a major and sensitive factor employed in many of the problems that are being investigated.

虽然一个问题可能会导致多种症状，但在研究中认为最主要、最敏感的因素是局部放电（PD）的成因。

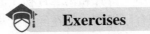 **Exercises**

Ⅰ. Choose the best answer into the black.

1. Overhead lines are the most ____ and frequently used carriers for electric energy.
   A. important   B. Cost effective   C. expensive   D. complicated

2. ____ sections of a conductor can also be repaired or replaced when damaged.
   A. Cheap   B. Long   C. Short   D. Too many

3. Over which electrical power is transmitted have been continuously increasing along with system ____ levels.
   A. voltage   B. current   C. power   D. circuit

Ⅱ. Answer the following questions according to the text.

1. What is the CM？

2. What is the PD ?

3. What are the most important and the most expensive components of EHV lines?

Ⅲ. Translate the following into Chinese.

Transmission line conductors are the most important and the most expensive components of EHV lines; they are, however, susceptible to ageing. One of the main reasons limiting their lifetime is unavoidable wind induced aeolian vibrations. Many important lines in several countries are now reaching an age where the mechanical safety tends to unacceptably low values, a condition which is very often little understood. Also, due to the increasing use of new types of conductors (e.g. new ACSR stranding, new materials, AAAC, ACAR, Alumoweld) where long term design experience does not exist, serious conductor fatigue and failures occur, causing increasing concern.

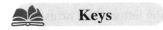

Keys

Ⅰ. Choose the best answer into the black.

1. B   2. C   3. A

Ⅱ. Answer the following questions according to the text.

1. The new methods to enable prompt identification of defective equipment under service conditions, i.e. through on-line processing for monitoring the health of equipment.

2. partial discharge.

3. The transmission line conductors.

Ⅲ. Translate the following into Chinese.

输电线路是超高压线路中最重要，也是最昂贵的一部分。但它们易老化，而老化的主要成因是风引起的振动。一些地区的重要线路已经达到使用年限，其机械安全性已经处于非常低的水平，但这种情况常易被人们忽视。虽然人们越来越多的使用新型导体材料（如新型钢芯铝绞线、新材料、铝合金绞线、铝合金芯铝绞线、铝包钢线等）的导线，但是由于缺乏长期的设计经验，这类导线易出现严重疲劳和故障的情况也逐渐被人们关注。

翻译技巧之状语从句（一）

英语状语从句包括表示时间、原因、条件、让步、目的等各种从句。我们已经谈到英语状语从句的一些译法。现在就一些比较常见的处理办法再进一步说明如下：

关于表示时间的状语从句介绍如下。

1. 译成相应的表示时间的状语

(1) While she spoke, the tears were running down.

她说话时，眼泪直流。

(2) When the history of the Nixon Administration is finally written, the chances are that Chinese policy will stand out as a model of common sense and good diplomacy.

当最后撰写尼克松政府的历史时，谈到对华政策可能成为懂得常识和处理外交的楷模。

以上两例译文中表示时间的状语置于句首，与原文一致。

(3) Please turn off the light when you leave the room.

离屋时请关电灯。

上句原文中表示时间的从句后置，译文中前置。

2. 译成"刚（一）……就……"的句式

(1) He had hardly rushed into the room when he shouted, "Fire! Fire!"

他刚跑进屋就大声喊："着火了！着火了！"

(2) Hardly had we arrived when it began to rain.

我们刚一到就下雨了。

(3) When I reached the beach, I collapsed.

我一游到海滩，就昏倒了。

(4) He had scarcely handed me the letter when he asked me to read it.

他把信一交给我，就叫我念给他听。

3. 译成并列的分句

(1) He shouted as he ran.

他一边跑，一边喊。

(2) They set him free when his rasom had not yet been paid.

他还没有交赎金，他们就把他释放了。

在上面两例的原文中，表时间的从句后置，在译文中提前。

(3) I was about to speak, when Mr. Smith cut in.

我正想讲，斯密斯先生就插话嘴了。

(4) Mrs.Acland gazed at him, her eyes darkening with a curious expressopm of dislike and distrust as he silently turned away.

阿克莱太太死瞪着他，眼神越来越阴沉，显示出一种对他又厌恶又怀疑的难以形容的表情，这时他只好默不作声地把脸转了过去。

在第（3）、（4）两例的原文里表时间的从句后置，在译文中不提前。

Section 2　Effect of Desert Environmental Conditions on The Flashover Voltage of Insulators

In Egypt, the insulators of overhead transmission lines and substations are often subjected to the deposition of contamination substances from the desert. This can lead to serious reduction in insulator effectiveness, resulting in flashovers and outages of electricity supply. It is important to mention that a remarkably high rate of interruption of 500kV and 220kV transmission lines, in Egypt, are recorded during spring seasons in desert areas where occasional sandstorms occur (Khamassine). This interruption in the power system will lead to the delay of the development of the community.

The flashover characteristics are thoroughly investigated for porcelain insulators exposed to natural sandstorms, as well as to simulated sandstorms with and without charged grids.

Test results show that neither natural nor artificial sandstorms affect the fast flashover voltage if the sand particles are not charged, whereas charged particles of sands reduce the flashover voltage of the insulators. To a higher extent, this reduction in flashover voltage will be greater as the grid is charged with DC voltage.

1. Introduction

To a large extent, the type of environmental conditions will significantly affect the insulators of overhead transmission lines. The desert climate is characterized by sand storms or hurricanes which contain very high speed sand particles. These sand particles hit the surface of any material and cause some erosion of it. A sandstorm in a desert is an important factor, which decreases the reliability of the transmission line. The decrease in reliability could occur as a result of the conductor swinging, which could decrease the spacing between the phases, or due to the decrease in the specific breakdown voltage of the polluted air gaps.

Regarding the pollution of the insulators in the desert, it has been generally concluded that:

(1) The early morning dew in the desert represents a major source of wetting the insulators.

(2) Sand storms increase the pollution of the insulators severely, the worst conditions occurring when sand storms are accompanied or followed by high humidity, or by rainy or misty weather.

(3) Pollution layers accumulated on insulators during sandstorms can be of larger grain size and have higher salt content than those accumulated under normal desert weather. The sandstorm pollution is usually carried by strong winds from distant regions.

(4) The effect of desert climatic conditions on the flashover voltage of conventional insulators is thoroughly investigated using a simulator of sandstorms.

2. Experimental set up and techniques

A string of porcelain insulators, as shown in Fig.6.1, but with five units, is selected for this study, each unit having 55cm leakage path, 15cm suspension length and 33cm shed diameter. The string is energized with AC voltage and suspended 1.5m above the set up device, which blows a quantity of sand, completely controlled by a shutter. The sand is blown by using four cylindrical turbines, which storm the air, carrying the sand upwards on the energized insulator string.

The sand grain size is selected to be less than 250μm using multi-sieves, and the velocity of the sand storming air onto the insulator is 10 m/s. The set up device used for the experimental work to simulate the artificial sand storm is shown in Fig.6.2.

Fig.6.1　The insulator string

3. Results and discussion

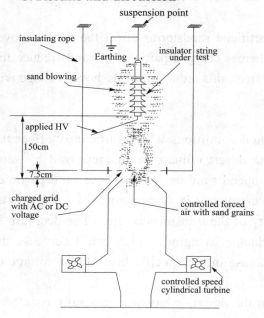

Fig.6.2 The set up device used for artificial sand storm

(1) Effect of the blown sandon the flashover voltage of the insulator string.

The insulator string is suspended vertically, whereas the storming sand is blown upwards from the set up device, so as to cover the string completely. Subsequently, the AC high voltage is applied on the insulators gradually until flashover voltage occurs. The same procedure is repeated using natural sand storms horizontally, with a sand speed of 11m/s. flashover voltage values of the string with storming air, as well as with blown sand (both artificial and natural), are shown in Fig.6.3.

The natural case of study was measured during natural sand storms, which usually occur in Egypt during Khamassine. The measuring was taken during these natural storms and recorded for the purpose of comparison in this study. The artificial studies are those studied using the set up device mentioned to simulate the sand storm.

Fig.6.3 shows the flashover voltage of the string under different conditions, with only storming air of 10m/s speed [representing wind without sand (case A)], with artificial sand storm of 10m/s speed with uncharged sand (case B) and, finally, for natural sand storm of 11m/s speed [measured and recorded during a natural and actual sand storm (case C)].

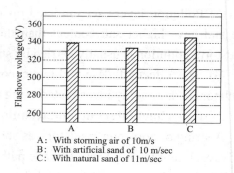

Fig.6.3 Flashover voltage of string insulator with and without blown sand (natural and artificial sandstrom)

From the results shown in Fig.6.3, it can be deduced that to a lesser degree, the uncharged sand particles reduce the flashover voltage of the insulators, whereas the natural sand storms, being horizontal at a fast speed, dissipate the formed arc along the string and cause higher flashover voltage of the string (case C).

(2) Effect of charged grid on the flas- hover voltage of insulators

Occasional sand storms (Khamassine) represent the main problem of interruption of high voltage overhead transmission lines. This occurs when sand particles are electrically charged due to their long exposure to an electric field, resulting in covering the whole string with charged particles after which a complete flashover occurs.

For simulating the effect of the charged particles on the flashover voltage, a charged grid is placed 7.5cm apart from the sandstorm set up device. The grid is electrically energized with AC voltage with power frequency 50Hz with values of 3, 5kV and 7kV, respectively. The test results are shown in Fig.6.4. Finally, the experiment has been repeated with applying a DC voltage of 5kV on the grid. For both conditions (applying AC and DC voltages to the grid), the AC voltage applied to the insulator string under test was gradually raised until flashover voltage occurred while the sand is blowing on the string. Fig.6.5 shows a comparison of the results obtained for the different conditions of study, namely with and without uncharged sand blowing on the insulator and with sand blowing while the grid is charged with 5kV AC and, finally, with the grid charged with 5kV DC.

From the test results shown in Fig.6.4, it can be deduced that the flashover voltage of the string decreases as the grid voltage increases. However, as expected, the reduction in flashover voltage is significantly higher as the grid is charged with DC voltage, as shown in Fig.6.5.

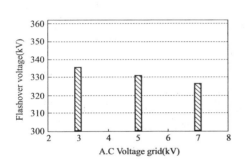

Fig.6.4　Flashover voltage of the string under the effect of the charged grid of AC voltage

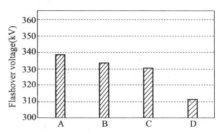

Fig.6.5　Effect of charging grid with AC and DC voltages on flashover voltages

The results obtained show that in the case of applying AC voltage with power frequency of 50Hz on the grid, the flashover voltage of the insulator is decreased. Moreover, it is found that the flashover voltage of the insulator is lower for the case of applying a DC voltage to the grid compared with an AC voltage having the same value of voltage. This is because the effective value of AC is lower than that of the DC value [the effective value of the AC voltage= $1/\sqrt{2}\, E_{max}$. (in the case of a sinusoidal wave)].

4. Conclusion

To a large extent, insulators show a significant change in electrical performance when exposed to desert environmental conditions. Either natural or artificial sandstorms affect fast flashover voltage. Charged sand particles reduce the flashover voltage of insulators. To a higher extent, this reduction in flashover voltage will be greater as the grid is charged with DC voltage.

第二节　沙漠环境对绝缘体闪络电压的影响

在埃及，架空输电线路和变电站的绝缘子经常受到来自沙漠沉积物质的污染。这可能会显著地降低绝缘子的有效性，从而使线路易发生闪络和断电。值得注意的是，在 500kV 和 220kV 输电线路中这种断电现象较易发生。有记录显示，在埃及，这种现象常发生于春季偶尔有沙尘暴的沙漠地区（哈马斯地区）。而电力系统断电会使社会发展受阻。

利用通电或是不通电的电网来模拟沙尘暴，对暴露在自然沙尘暴环境中的陶瓷绝缘子进行了闪络特性的研究。

试验结果表明：如果沙粒不带电，那么无论是自然还是人工沙尘暴都不会影响闪络电压，然而沙粒带电，则会降低绝缘子的闪络电压。进一步来说，如果电网带有直流电，闪络电压将会降低更多。

1. 引言

环境条件将在很大程度上影响架空输电线路上的绝缘子性能。含有高速沙粒的沙尘暴和飓风是沙漠气候特点之一。这些细小的沙粒会击打各种材料的表面，同时对其造成侵蚀。沙漠中沙尘暴是造成输电线路不稳定的一个重要因素。其可靠性降低的主要原因有两方面：导线摆动造成的导线相间距离降低和气隙污染造成的故障电压降低。

关于沙漠中绝缘子的污染，一般可以归纳为以下几点：

（1）清晨的露水是导致绝缘子潮湿的主要原因。

（2）沙尘暴会加剧绝缘子的污染，而伴随着高湿度、大雨或大雾的沙尘暴影响更大。

（3）由于沙尘暴污染通常由大风从遥远的地区携带过来，因此，相较于正常沙漠天气，沙尘暴天气下绝缘子积累的污染层的晶粒尺寸更大且含盐量更高。

（4）使用沙尘暴的模拟器可以彻底研究沙漠气候条件对常规绝缘子的闪络电压的影响。

2. 实验装置和实验方法

图 6.1　绝缘子串

这次被选来进行研究的是一个陶瓷绝缘子串，如图 6.1 所示，它由五片绝缘子组成。每片绝缘子放电路径长 55cm，悬浮长度 15cm，伞径 33cm。绝缘子串由交流电压供电并位于实验装置上方 1.5m，该实验装置完全由快门控制吹沙。通过一个百叶窗来控制风向，并利用四个柱状涡轮机鼓风将沙朝上吹到绝缘子串上。

选取尺寸小于 250μm 的沙粒，打在绝缘子上沙尘暴的风速为 10m/s。模拟人工沙尘暴的实验设备示意图如图 6.2 所示。

3. 结果和讨论

（1）扬沙对绝缘子串闪络电压的影响。

绝缘子串竖直悬挂，运行吹沙装置使沙自下而上完全包裹绝缘子。随后在绝缘子上加载交流高压电直到绝缘子串发生闪络。随后缓慢增加绝缘子串上的交流高压，直至达到击穿电压。在自然沙尘暴环境下水平地重复上述步骤，其中沙粒速度为 11m/s。风暴、人工沙尘暴及自然沙尘暴下的闪络电压值如图 6.3 所示。

研究的自然案例数据是在天然沙尘暴期间测量得到的，这种天气常在埃及的哈马斯地区发生。在这些自然风暴期间对其进行测量并记录，作为本研究的对比。而人工研究则是利用前面提到的模拟沙尘暴的装置进行的。

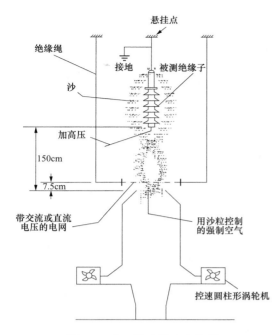

图 6.2 人工沙尘暴产生装置

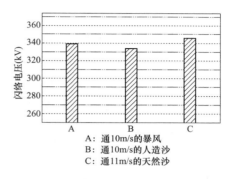

图 6.3 各种情况下闪络电压值

图 6.3 给出了不同条件下的闪络电压，这些条件包括速度为 10m/s 的暴风［代表不含沙粒的风（情况 A）］、速度为 10m/s 的含不带电沙粒的人工沙尘暴（情况 B），和速度为 11m/s 的自然沙尘暴［在自然和实际的沙尘暴（情况 C）期间测量和记录的］。

从图 6.3 可以得出，不带电荷的沙粒较小程度地降低了绝缘子闪络电压，而自然条件下的沙尘暴做高速度水平运动，会驱散绝缘子串上的电弧并造成更高电压的绝缘子闪络（情况 C）。

（2）电网对绝缘子闪络电压的影响。

偶发的沙尘暴（哈马斯地区）是输电线路断电的主要问题。这是因为沙粒带电并长时间暴露于电场中，最终会致使带电粒子覆盖整个绝缘子表面而发生闪络。

为了模拟闪络电压中带电粒子的影响，将一个带电电网安装在离沙尘暴发生装置 7.5cm 处，该电网频率为 50Hz，依次带有 3、5kV 和 7kV 的交流高压电。实验结果如图 6.4 所示。最后，将 5kV 的直流电压加于电网上，重复该试验。在两种情况（将直流和交流电压分别加于电网上）下，将沙粒吹向绝缘子串的同时，使施加在绝缘子串上的交流电压逐渐上升到闪络电压。两种情况的结果对比，如图 6.5 所示，即将有无带电沙粒吹向绝缘子，而沙粒分别利用带 5kV 直流或交流电的电网进行充电。

从图 6.4 可以看出，绝缘子串闪络电压随电网电压增加而增加。但正如预期，减少的闪络电压远大于电网上的直流电压，如图 6.5 所示。

结果表明，在电网上施加 50Hz 的交流电时，绝缘子的闪络电压会降低。此外，与施加相同电压值的交流电相比，向电网施加直流时绝缘子串的闪络电压较低。这是由于交流的有效值比直流的低［交流电压得有效值为直流电压有效值的 $1/\sqrt{2}$（正弦波情况下）］的缘故。

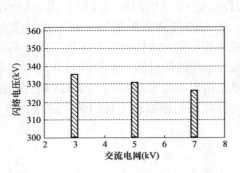 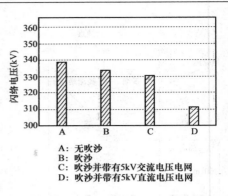

图6.4 交流电网作用下绝缘子串闪络电压　　图6.5 交、直流电网对闪络电压的影响

4. 结论

在很大程度上而言，暴露在沙漠环境中的绝缘子电气性能出现了显著变化。自然和人造沙尘暴都会影响闪络电压，带电荷的沙粒会降低绝缘子的闪络电压。进一步来说，如果电网通直流电压，这种降低会更大。

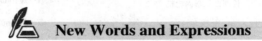

New Words and Expressions

units	n. 机组
shed diameter	n. 伞径
turbines	n. 涡轮
insulator string	绝缘子串
suspension length	悬空长度

Notes

1. The decrease in reliability could occur as a result of the conductor swinging, which could decrease the spacing between the phases, or due to the decrease in the specific breakdown voltage of the polluted air gaps.

沙漠中沙尘暴是造成输电线路不稳定的一个重要因素。其可靠性降低的主要原因有两方面：导线摆动造成的导线相间距离降低和气隙污染造成的故障电压降低。

2. The string is energized with AC voltage and suspended 1.5m above the set up device, which blows a quantity of sand, completely controlled by a shutter. The sand is blown by using four cylindrical turbines, which storm the air, carrying the sand upwards on the energized insulator string.

绝缘子串由交流电压供电并位于实验装置上方1.5m，该实验装置完全由快门控制吹沙。通过一个百叶窗来控制风向，并利用四个柱状涡轮机鼓风将沙朝上吹到绝缘子串上。

3. Fig.6.3 shows the flashover voltage of the string under different conditions, with only storming air of 10m/s speed [representing wind without sand (case A)], with artificial sand storm of 10m/s speed with uncharged sand (case B) and, finally, for natural sand storm of 11m/s

speed [measured and recorded during a natural and actual sand storm (case C)].

图 6.3 给出了不同条件下的闪络电压，这些条件包括速度为 10m/s 的暴风［代表不含沙粒的风（情况 A）］、速度为 10m/s 的含不带电沙粒的人工沙尘暴（情况 B）和速度为 11m/s 的自然沙尘暴［在自然和实际的沙尘暴（情况 C）期间测量和记录的］。

4. Test results show that neither natural nor artificial sandstorms affect the fast flashover voltage if the sand particles are not charged, whereas charged particles of sands reduce the flashover voltage of the insulators. To a higher extent, this reduction in flashover voltage will be greater as the grid is charged with DC voltage.

试验结果表明：如果沙粒不带电，那么无论是自然的还是人工沙尘暴都不会影响闪络电压，然而沙粒带电，则会降低绝缘子的闪络电压。进一步来说，如果电网带有直流电，闪络电压将会降低更多。

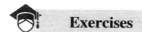

Exercises

Ⅰ. Choose the best answer into the black.

1. This can____to serious reduction in insulator effectiveness.
   A. lead   B. can   C. transport   D. own

2. To a large extent, the____of environmental conditions will significantly affect the insulators of overhead transmission lines.
   A. importance B. mean   C. type   D. sand

3. The early____dew in the desert represents a major source of wetting the insulators.
   A. afternoon   B. morning   C. night   D. evening

Ⅱ. Answer the following questions according to the text.

1. What are the reasons of the pollution of the insulators in the desert?
2. What are the reasons of the decreases the reliability of the transmission line?
3. What the device we use in this experiment?

Ⅲ. Translate the following into Chinese.

The results obtained show that in the case of applying AC voltage with power frequency of 50Hz on the grid, the flashover voltage of the insulator is decreased. Moreover, it is found that the flashover voltage of the insulator is lower for the case of applying a DC voltage to the grid compared with an AC voltage having the same value of voltage. This is because the effective value of AC is lower than that of the DC value [the effective value of the AC voltage= $1/\sqrt{2}\ E_{max}$. (in the case of a sinusoidal wave)].

Keys

Ⅰ. Choose the best answer into the black.

1. A   2. C   3. B

Ⅱ. Answer the following questions according to the text.

1. The early morning dew，Sand storms and Pollution layers.

2. The sandstorm.
3. A string of porcelain insulators.
III. Translate the following into Chinese.

结果表明，在电网上施加 50Hz 的交流电时，绝缘子的闪络电压会降低。此外，与施加相同电压值的交流电相比，向电网施加直流时绝缘子串的闪络电压较低。这是由于交流的有效值比直流的低［交流电压得有效值为直流电压有效值的 $1/\sqrt{2}$（正弦波情况下）］的缘故。

翻译技巧之状语从句（二）

关于表示原因的状语从句介绍如下。

1. 译成表示"因"的分句

(1) The crops failed because the season was dry.

因为气候干燥，作物歉收。

(2) Because we are both prepared to proceed on the basis of quality and mutual respect, we meet at a moment when we can make peaceful cooperation a reality.

由于我们双方都准备在平等互尊的基础上行事，我们在这个时刻会晤就能够使和平合作成为现实。

(3) The policies open to developing countries are more limited than for industrialized nations because the poorer ecomonies respond less to changing conditions and administrative control.

由于贫困国家的经济对形势变化的适应能力差一些，政府对这种经济的控制作用也小一些，所以发展中国家所能采取的政策比工业化国家就更有局限性。

"由于""因为"是汉语中常常用来表"因"的关联词。一般来说，汉语表示"因"分句在表示"果"之前，英语则比较灵活。但在现代汉语中，受西方语言的影响，也有放在后面的，如：

(4) She could get away with anything, because she looked such a baby.

她怎么捣蛋都没事，因为她看上去简直还像个娃娃模样。

(5) The strike leaders were alarmed when I told him what had happened as the reporter was unfriendly.

当我把发生的事情讲给罢工领导人听时，他们吓了一跳，因为这个记者是不友好的。

2. 译成因果偏正复句中的主语

(1) Because he was convinced of the accuracy of this fact, he stuck to his opinion.

他深信这件事正确可靠，因此坚持己见。

(2) Pure iron is not used in industry because it is too soft.

纯铁太软，所以不用在工业上。

(3) The perspiration embarrasses him slightly because the dampness on his brow and chin makes him look more tense than he really is.

额头和下巴上出的汗使他看起来比他实际上更加紧张，因而出汗常使他感到有点困窘。

3. 译成不用关联词而因果关系内含的分句

(1) "You took me because I was useful. There is no question of gratitude between us."said Rebecca.

"我有用，你才收留我。咱们之间谈不到感恩不感恩。"丽贝卡说。

(2) After all, it did not matter much, because in 24 hours, they were going to be free.

反正关系不大，二十四小时以后他们就要自由了。

(3) He was not only surprised but, to start with, extremely suspicious, as he had every reason to be.

他不但惊讶而且首先是十分怀疑，他这样感觉是完全有理由的。

Section 3　Failures in Outdoor Insulation Caused by Bird Excrement

Outdoor insulation in electrical systems is usually contaminated by the action of external agents around it. Contamination can be either natural or industrial. The climate, soil type, as well as the proximity of the coasts cans enerate natural contamination. Another source of polluting agents, considered natural, is bird excrement (BE). This contamination, combined with humidity, can cause flashovers on outdoor insulation, mainly in transmission lines.

The problem of contamination by BE has not a unique solution, since a diversity of variables must be considered, when dealing with it, such as bird type, excrement type, climate, physical structures, etc. Several programs and methods have been used to reduce the impact of BE contamination, in different utilities. The time of application of these programs is significant when the objective is to avoid the birds' settling in the site. A suitable action reduces the time and effort required to impel them away. For example, the use of scaring devices is more effective at the beginning of the migration cycle. Additional factors to consider are the application period, the planning, and the organization of the selected program. Sometimes it is necessary to apply more than one methodology due to birds' variety. If a program is not producing the desired effects, it is essential to analyze the causes and possible modifications before rejecting it. The key to success in birds' dispersion programs are time, persistence, organization, and miscellany.

1. Flashover mechanism by BE

The BE is a mixture of urine and solid lees. The BE is classified according to three forms: rod or bar, snail, and fluid. The last one is produced by birds of prey and herons. Carnivorous birds, which consume great amounts of proteomics weaves, produce great volumes of urethral liquid. This provokes a white or light (cream) color in their excrement. Factors that can modify the volume and consistency of the BE are: species, age, lunch time, dietetic content, amount of food and water consumption, females' reproductive cycle, and diseases. Birds can defecate at any time, but some do not when asleep, in which case volume of excrement in the morning is abundant. It is common for the birds to eliminate the excrement before beginning their daily activities.

Big birds' fluid excrement liberation can cause line outage in continuous and conductive lines, and even a short circuit between the structure and the conductor. If the excrement release occurs near an insulator in vertical position, the arc distance of the insulator could short circuit in similar way.

2. Anti-bird control systems

The selection of an anti-bird device depends on the bird type (bird size, way of life, etc.), and the grounds for the birds to settle on the structure (rest, nesting, protection, etc.) Several companies offer different products to avoid the incursion of birds. By their function, these products are classified in eliminators of birds, dissuasion devices, physical barriers, and covering devices.

3. Physical barriers

The use of physical barriers depends on the kind of installation to be protected. Distribution lines are in general more critical than transmission lines because distance between phases is more reduced in the first ones, which could be affected only by the presence of big birds. On the other hand, the big space between phases in the case of transmission lines makes all kind of birds to fly or even settle between the two phases. In distribution lines, the objective is to avoid the birds flying or settling between the two phases to avoid the bird electrocution, and a subsequent short circuit. On the other hand, in transmission lines, where only eagles or similar birds could cause a short circuit, physical barriers are used mainly to avoid the birds perch stand, and hence, to reduce contamination by BE.

There exist two main types of physical barriers: tips and meshed.

4. Tips protectors

Tips protectors are peaks placed on the cross arms where the birds inhabit. Three parameters must be considered in the selection of these devices: the peaks size, the separation between peaks, and the peaks' forms. If the peaks are too separated, the tips will lose effectiveness for small birds but they will protect a larger area against big birds. On the other hand, if the peaks are very long, big birds can settle on the peaks. If metallic tips are used, additional problems come out. Consequently, tips must be made up of insulating material.

The main disadvantage of tips protectors is that they become an impediment for maintenance activities. Due to this, one of the requirements to select tips protector is their easy to be removed. This requirement represents another problem, because if tips protector is not well fastened to the structure, it could fall down by wind action. On the other hand, if its installation is too rigid, it will consume more timework to be removed. The useful time of tips protector is quite long, and depends on the material and the mechanical resistance of the device (Fig.6.6).

5. Meshed

Meshed consists of a net or threads placed on the area to be protected. Their efficiency is good in specific areas. However, their use is not feasible in areas with strong wind or many bird specimens. The dimension of the mesh must be selected according to bird size, in order to avoid the entrance of small species in the protected area. The disadvantage of this kind of barrier is that, by wind action, solid residues are thrown onto the mesh. This rubbish will be holding on the mesh requiring periodic maintenance. Besides, this solutio aggressive towards the fauna, because several animals are caught into the mesh (birds, bats, etc.). The lifetime of this solution can be from 1 year up to 15 years with reasonable maintenance.

Fig.6.6 Tips protector in a transmission line

 原文翻译

第三节　鸟粪对户外绝缘的影响

电力系统户外绝缘子经常会受到来自周边环境的污染，而污染可能来源于大自然或是工业。气候条件、土壤类型或者临近海洋等因素都会产生自然污染，而另一种被认为是自然污染的污染源称作鸟粪（BE）。这种带有一定湿度的污染物，主要会使输电线路上的户外绝缘子发生绝缘闪络。

由于鸟粪具有多样性，就必须考虑鸟的种类、鸟粪种类、气候条件、物理结构等，因此，由鸟粪引起的绝缘子污染问题还没有一项有效的解决措施。在一些电力公司，已经使用了某些装置和方法来减少鸟粪对绝缘子污染的影响。为了避免鸟类在现场排泄粪便，这些解决措施的动作时间是十分重要的，而合适的动作能够有效地节省时间和劳动强度以便驱赶鸟类。例如，在鸟类迁徙周期开始时刻，采用恐吓装置的效果非常显著。此外，还应该考虑装置的一些其他因素，包括投放时期、计划及组织。由于鸟类的多样性，有时候非常有必要采取多种方法以驱赶鸟类。如果某些方法没有达到预期设计的效果，那么在否定结论之前，十分有必要分析该结果的产生原因并进行改进。驱散鸟类成功的关键要素是：合适的时间、耐心、计划性和多方法结合。

1. 鸟粪引起的闪络机理

鸟粪是尿液和固体残渣组成的混合物。根据粪便形状，分为三种形式：杆式或棒式、螺旋式、流体式，而第三类是由猛禽或者苍鹭产生。食肉鸟类会消耗大量的蛋白质，继而产生大量尿液，这在鸟粪中以乳白色或浅色呈现。鸟粪的体积大小和黏稠度的影响因素包括：鸟的种类、年龄、进食时间、食物组成、食物数量和水耗量、雌性繁殖周期及疾病等。鸟类可以随时进行排泄，不过有些鸟类栖息时不排泄，所以这些鸟类在早晨的排泄量很大。

不过，鸟类在开始日常活动之前的排便是一种正常的、普遍的行为。

大型鸟类的液体排泄物会引起输电线路供电中断，甚至导致杆塔与导线之间发生短路故障。如果排泄物的位置靠近悬垂绝缘子外侧，将会在绝缘子的空气间隙之间产生类似的电弧短路。

2. 驱鸟装置

驱鸟装置的选择取决于鸟的类型（鸟类尺寸、生活习性等）和鸟类在输电线路上的停留位置（包括栖息、筑巢、防护等）。几家公司专门提供了不同的防鸟措施以阻止鸟类的入侵，根据设备的功能不同，可分为防鸟装置、驱鸟装置、物理保护和遮蔽装置。

3. 物理保护

物理保护取决于安装物的种类。一般地，配电线路比输电线路的安装要求更苛刻。一方面是配电线路上的相线之间的几何间距更小，这只有大型的鸟类才能产生影响，而另一方面，输电线路相线的几何距离大可以使各种鸟类顺利飞过，甚至栖息。在配电线上，物理保护的目标是避免鸟类在相线之间飞翔或是停留，以便避免鸟类触电而导致的短路故障；输电线路上，只有鹰或类似的鸟类才可能导致短路故障，而物理保护措施主要用于阻止鸟类停留，以便减少鸟粪引起的绝缘子污染（BE）。

物理保护措施主要有 2 种主要类型：防鸟刺和防鸟罩。

4. 防鸟刺

防鸟刺是一个辐射状的金属刺群，安装在鸟类习惯停留的铁塔横担上。防鸟刺装置的设计必须考虑三个因素：鸟刺的尺寸、鸟刺的几何间距、鸟刺的类型。如果刺与刺的几何间距过大，那么将会失去对小型鸟类的防护效果，但能有效地阻止大型鸟类的入侵。另外，如果刺的长度过大，那么大型鸟类就能停留在上面。如果使用金属质防鸟刺，也会带来其他的问题，因此，防鸟刺必须采用绝缘材料制造。

防鸟刺的主要缺点是对输电线路的运行维护产生障碍，基于这个缺陷，选择防鸟刺的要求之一是拆除的方便性。然而，这也导致另外一个问题的出现，如果防鸟刺安装不牢靠，那么它会由于风力作用而掉落。此外，如果防鸟刺安装太过牢靠，那么将会花费更多时间进行拆除。防鸟刺的有效工作的时间长，短取决于装置的制造材料和机械性能（图 6.6）。

5. 防鸟罩

防鸟罩，由安装在被保护区域的金属网或是线组成。在某些区域，其防护效果非常理想。然而，在有强风或鸟类密集地区，使用防鸟罩是不可行的。为了避免小型鸟类进入被保护区域，那么需要根据鸟的大小设计合适的网孔尺寸。防鸟罩的缺点是：会受到大风及固体残留物的影响，

图 6.6　输电线路上的防鸟刺

因此，防护网上的排泄物需要定期进行清理。此外，由于有些动物会陷入到网孔里面（如鸟类、蝙蝠等），因此，防鸟罩不利于保护野生动物。如果对防鸟罩进行合理的运行维护，那么

该装置可正常地运行 1 年至 15 年。

 New Words and Expressions

impel	v. 驱赶，推动
electrocution	n. 触电死亡
eliminators	n. 整流器
impediment	n. 障碍物，阻碍

 Notes

1. The problem of contamination by BE has not a unique solution, since a diversity of variables must be considered, when dealing with it, such as bird type, excrement type, climate, physical structures, etc.

由于鸟粪具有多样性，就必须考虑鸟的种类、鸟粪种类、气候条件、物理结构等，因此，由鸟粪引起的绝缘子污染问题还没有一项有效的解决措施。

2. Factors that can modify the volume and consistency of the BE are: species, age, lunch time, dietetic content, amount of food and water consumption, females' reproductive cycle, and diseases. Birds can defecate at any time, but some do not when asleep, in which case volume of excrement in the morning is abundant.

鸟粪的体积大小和黏稠度的影响因素包括：鸟的种类、年龄、进食时间、食物组成、食物数量和水耗量、雌性繁殖周期及疾病等。鸟类可以随时进行排泄，不过有些鸟类栖息时不排泄，所以这些鸟类在早晨的排泄量很大。

3. The main disadvantage of tips protectors is that they become an impediment for maintenance activities. Due to this, one of the requirements to select tips protector is their easy to be removed.

防鸟刺的主要缺点是对输电线路的运行维护产生障碍，基于这个缺陷，选择防鸟刺的要求之一是拆除的方便性。

 Exercises

I. Choose the best answer into the black.
1. Contamination can be either____ or industrial.
   A. important   B. dangerous   C. natural   D. artificial
2. The BE is a mixture of____ and solid lees.
   A. feed   B. nuts   C. urine   D. visceral
3. ____ consists of a net or threads placed on the area to be protected.
   A. Tips protectors   B. Item   C. People   D. Meshed

II. Answer the following questions according to the text.
1. What does depend on the selection of an anti-bird device?
2. What does depend on the use of physical barriers?

3. What must be considered in the selection of these devices?

Ⅲ. Translate the following into Chinese.

The use of physical barriers depends on the kind of installation to be protected. Distribution lines are in general more critical than transmission lines because distance between phases is more reduced in the first ones, which could be affected only by the presence of big birds. On the other hand, the big space between phases in the case of transmission lines makes all kind of birds to fly or even settle between the two phases. In distribution lines, the objective is to avoid the birds flying or settling between the two phases to avoid the bird electrocution, and a subsequent short circuit. On the other hand, in transmission lines, where only eagles or similar birds could cause a short circuit, physical barriers are used mainly to avoid the birds perch stand, and hence, to reduce contamination by BE.

Keys

Ⅰ. Choose the best answer into the black.

1. C   2. C   3. D

Ⅱ. Answer the following questions according to the text.

1. The bird type (bird size, way of life, etc.), and the grounds for the birds to settle on the structure (rest, nesting, protection, etc.).

2. The kind of installation to be protected.

3. The peaks size, the separation between peaks, and the peaks' forms.

Ⅲ. Translate the following into Chinese.

物理保护取决于安装物的种类。一般地，配电线路比输电线路的安装要求更苛刻。一方面是配电线路上的相线之间的几何间距更小，这只有大型的鸟类才能产生影响，而另一方面，输电线路相线的几何距离大可以使各种鸟类顺利飞过，甚至栖息。在配电线上，物理保护的目标是避免鸟类在相线之间飞翔或是停留，以便避免鸟类触电而导致的短路故障；输电线路上，只有鹰或类似的鸟类才可能导致短路故障，而物理保护措施主要用于阻止鸟类停留，以便减少鸟粪引起的绝缘子污染（BE）。

翻译技巧之状语从句（三）

关于表示条件状语的从句介绍如下。

1. 译成表"条件"的分句

(1) "Sure, there're jobs. There is even Egbert's job if you want it."

"当然，工作是有的。只要你肯干，甚至就可以顶埃格伯特的空缺。"

(2) It was better in case they were captured.

要是把他们捉到了，那就更好了。

(3) They can't go through any city! If they lay over, it's got to be in a special garage.

他们不能穿过任何城市！如果中途要停顿，就得进专门的车库。

(4) Once the men had been accepted by the Comet organization, they were brought to Brussels.

他们一旦被彗星组织收下了,就被带到布鲁塞尔去。

"只要""要是""如果""一旦"等都是汉语表示"条件"的常用关联词,在语气上,"只要"("只有")最强,"如果"最弱,"如果"也用来表示假设。英语中表示"条件"的从句前置后置比较灵活,汉语中表示"条件"的分句一般前置。

2. 译成表示"假设"的分句

(1) If one of them collapsed, as they often did, the guide used to carry him over the mountains.

如果其中一人垮了,这种事时常在他们中间发生,向导就要背着他过山。

(2) If the government survives the confidence vote, its next crucial test will come in a direct Bundestag vote on the treaties May 4.

假使政府经过信任投票而保全下来的话,它的下一个决定性的考验将是五月四日在联邦议院就条约举行的直接投票。

(3) If the negotiation between the rich northerly nations and the poor southerly nations make headway, it is intended that a ministerial session in December should be arragned.

要是北方富国和南方穷国之间的谈判获得进展的话,就打算在十二月份安排召开部长级会议。

"如果""只要""假设"等都是汉语中用来表示"假设"的常用关联词。汉语中表示"假设"的分句一般前置,但作为补充说明情况的分句则往往后置。

3. 译成补充说明情况的分句

(1) "He's dead on the job, Jesse. Last night if you want to know."

"他是在干活时死的,杰西。就是昨晚的事,如果你想知道的话。"

(2) "You'll have some money by then,— that is, if you last the week out, you fool."

"到那时你就该有点钱了——就是说,如果你能度过这个星期的话,你这傻瓜。"

(3) You can drive tonight if you are ready.

如果你准备好了的话,你今晚可以开车。

Chapter 7 Lightning Protection and Grounding

Section 1 Optimization of Hellenic Overhead High-voltage Transmission Lines Lightning Protection (1)

It is well known that the lightning protection of transmission lines is exclusively relying on their correct initial design. Although detailed engineering studies are usually performed by electric power utilities for the design of new transmission lines, there are reported cases where the design is based simply on tradition or on utilities' standardization policy. In this paper, the lightning protection of high-voltage transmission lines is faced as an optimization problem where optimum design parameters are calculated for the lines, relating their cost with the lightning failures' cost, aiming to reduce or even eliminate lightning failures. The optimization method considers all the available protection means, i. e. ground wires and surge arresters. In order to validate the effectiveness of the proposed method, it is applied on several operating Hellenic transmission lines of 150kV carefully selected among others due to their high failure rates during lightning thunderstorms. The obtained optimum parameters, which reduce the failure rates caused by lightning are compared with the operating transmission lines' existing parameters showing the usefulness of the method, which can prove to be a valuable tool for the studies of electric power system designers.

Security and reliability of power supply are very significant issues for energy utilities. Supply interruptions are mainly related to faults in overhead high-voltage transmission lines, due to lightning strokes on them, producing dangerous overvoltages and damages on equipments. The correct and careful design constitutes the most important issue for transmission lines, since slight differences in the design parameter values can affect significantly their lightning performance, determining the operation and the performance of transmission lines for the rest of their life.

Several studies have been conducted in an effort to either determine the minimum cost design of transmission lines, representing the total cost of the system as a function of the system design variables, or to explore the sensitivity of the required present worth of revenue, to several design variables, in order for design performance at minimum cost to be achieved. Furthermore, a range of line optimization techniques have been introduced, which can be applied to decide whether standard or optimized line designs are appropriate, while optimal design methods, for overhead high-voltage transmission lines, with the main objective being the minimization of the line total annual cost, considering the relevant technical constraints and both fixed and running cost items, have been presented. The economical aspects of the overhead distribution line lightning performance, taking into account the customer and utility costs of line outages, have been analyzed, while alternative design procedures for uncompensated overhead transmission lines based on the derivation of closed-form analytical expressions for both line power and current ratings, in terms of the geometrical data of the line tower and its bundled conductors, were introduced. Finally, an optimal design method for improving the lightning performance of overhead high-voltage transmission lines was proposed using a quasi-Newton optimization algorithm.

The current work inspired by the work presents in detail a new design methodology for the more effective lightning protection of transmission lines which incorporates all the available protection means. The proposed methodology in contrast to the work presented, which only selects optimum values for the line insulation level and the tower footing resistance, includes also the optimum selection of the energy absorption capability of surge arresters and the surge arresters' installation interval. The proposed methodology divides transmission lines into regions and separate designs are performed for each region, serving with best way lines, which are running at the same time through plain regions, coastlines and/or mountainous regions with significantly different meteorological and geographical characteristics through their length. Suitable performance indices are defined in order to relate the insulation level, the tower footing resistance, the energy absorption capability of the surge arrester and the surge arresters' installation interval cost values to the lightning failure costs. Using a more sophisticated optimization algorithm, in contrast to this, based on direct computation, optimum values of those four parameters are calculated in order to minimize the defined performance indices. It must be mentioned that the optimization methodology incorporates all the available protection means, i.e., ground wires and surge arresters, something which is not a common practice for all electric utilities' designs.

The developed methodology is applied on several operating Hellenic transmission lines (including open loop lines) of 150kV with high lightning failure rates, in order to validate its effectiveness. New values for the studied transmission line parameters are proposed, having as a result the efficient lightning protection of the lines. The proposed methodology can be particularly useful to the studies of transmission lines' designer engineers contributing effectively in the reduction and/or elimination of lightning failures.

1. Transmission lines' lightning failures

In order to protect the transmission lines and improve their lightning performance, overhead ground wires and surge arresters are used. The purpose of these protection methods is to reduce the insulations flashovers of the line that can cause power supply interruptions. For properly designed lines, most lightning strokes to the line are expected to terminate on the ground wires, but for high lightning currents and tower footing resistances back flashovers can occur. Moreover for any line, there is also the possibility of a shielding failure, which can lead to insulation flashover or a surge arrester failure when such protective devices are installed.

The total lightning failure of a transmission line N_T expressed in flashovers per 100km per year is the summation of arrester total failurerate A_T, shielding failurerate N_{SF} and back flashover failure rate N_{BF}. Thus

$$N_T = A_T + N_{SF} + N_{BF} \tag{7.1}$$

2. Surge arresters' failure rate

A surge arrester presents a momentary path to earth, which removes the superfluous charge from the line. The most common types of arresters are the open spark gaps, the SiC arresters with spark gaps and the metal oxide surge arresters without gaps. The last type, which is composed of non-linear resistors of metal oxide, mainly ZnO without spark gaps, is most commonly used today.

The main characteristics of a surge arrester are

(1) Continuous operating voltage (U_c): Designated rms value of power frequency voltage that may be applied continuously between the terminals of the arrester.

(2) Rated voltage: Maximum permissible rms value of power frequency voltage between arrester terminals at which it is designed to operate correctly under temporary over voltages.

(3) Residual voltage (U_{res}): Peak value of the voltage that appears between arrester terminals when a discharge current is injected.

(4) Rated discharge current: Peak value of lightning current impulse, which is used to classify an arrester.

(5) Energy absorption capability: Maximum level of energy injected into the arrester at which it can still cool back down to its normal operating temperature. Standards do not define energy capability of an arrester.

In IEC there exists the term line discharge class, but since this in not enough information, various manufacturers present thermal energy absorption capability in kJ/kV(U_c), defined as the maximum permissible energy that an arrester may be subjected to two impulses according to IEC clause8.5.5, without damage and without loss of thermal stability.

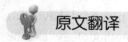
原文翻译

第一节　希腊高压架空输电线路防雷保护的最优化（1）

众所周知，输电线路雷击防护效果主要取决于正确的前期设计。虽然电力部门对新建输电线路进行了详细研究，然而有报告表明：这些研究都是基于传统的或是电力企业制定的标准规范。本文将高压架空输电线路的防雷保护视为最优化问题。基于最优化的输电线路设计参数，再结合雷击线路造成的损失，以减少甚至消除雷电事故。这种最优化方法考虑了各种切实可行的防护措施，例如，架设避雷线和安装避雷器。为验证该方法的有效性，从希腊正常运行的输电线路中，严格地选择了雷电天气下故障率较高的某些150kV电压等级输电线路用于展开研究。研究结果表明，最优化参数设计的输电线路比其他正常运行线路的雷击故障率更小，因此，本节的方法确实有效，这也给电力系统的防雷保护研究提供了一定的参考价值。

对于电网公司而言，安全性和可靠性是评价输电线路重要指标。雷击高压架空输电线路产生的电力设备过电压或机械性损坏，是导致电力系统供电中断的主要原因。输电线路最重要的一个环节是精准的参数设计，而细微的参数变化将显著地影响输电线路的防雷效果，并决定输电线路后续的运行状态和工作性能。

目前，所有研究工作都是为获得使成本费用最小化的参数设计值，一方面是为了确定输电线路设计的最小经济成本，即与设计参数有关的数学函数来描述电力系统的总费用；另一方面，是为了探索当前经济成本的敏感因素。此外，还介绍了一系列的输电线路优化技术，用于判别标准化或最优化的线路设计是否合理。在考虑相关技术限制、固定成本与运行成本的情况下，为使输电线路年运行总费用最小，提出了高压架空输电线路的最优化设计方案。该方案考虑了因供电中断对用户与工企业造成的经济损失，用于分析架空配电线路防雷性能的经济性。本节还阐述了一种无需考虑功率补偿的架空输电线路求解程序，同时，程序的求解公式是在传输功率和额定电流已知的基础上进行推导的，并且数据来源于输电线路杆塔和分裂导线的几何数据。最后，基于拟牛顿优化算法，提出了一种用于提高高压架空输电线路防雷性能的最优化求解算法。

受此启发，本文提出了一种更有效的输电线路雷击防护方法，但它并不包含所有切实可行的防护措施。相比于当前的防护措施，该方法只是选择了最优化的输电线路绝缘水平和杆塔接地电阻，包括最优的避雷器通流容量和间隙安装。该方法将输电线路按照地域不同进行划分，不同地域的输电线路分别进行设计，因为不同地域拥有不同的气候和地理环境，根据线路的路径长度，使输电线路同时运行于平原地区、沿海地区或山区环境，以使输电线路路径最优。为了提出合适的防雷性能评价指标，将绝缘水平、杆塔接地电阻、避雷器通流容量及避雷器安装间隙的成本费用和雷害损失费用联系在一起。与此相反，基于直接求解原理，提出了一种更加复杂的最优化算法，计算出四个参数的最优值，以使防雷性能评价指标最小化。值得注意的是，由于该方法并未包含所有切实可行的防雷保护措施，例如避雷线或避雷器，因此并不适用于所有电力院的防雷设计。

将新方法应用到希腊一些雷击故障率较高的 150kV 电压等级输电线路上（包括开环运行的输电线路），以验证其有效性，而运行结果表明，改进的输电线路设计参数有助于输电线路防雷保护。本节提出的方法有助于设计人员关于输电线路的研究工作，以更有效地减少或消除输电线路上的雷击故障。

1. 输电线路雷击故障

一般地，通过架设避雷线和安装避雷器来保护输电线路和提高线路的防雷性能。采用这些措施，可以减少由输电线路绝缘子闪络导致的供电中断时间。合理的输电线路设计，可以在避雷线末端消除大部分的雷击影响，而当雷电流过大和杆塔接地电阻过高时，则会出现反击现象。更有甚者，对于任意输电线路，当上述措施出现故障时，也会造成线路绝缘子闪络和避雷器防护性能失效。

输电线路的总雷击故障率 N_T，指每 100km 的线路每年因雷击引起的闪络次数，是避雷器失效率 A_T、保护失效率 N_{SF} 和反击闪络失效率 N_{BF} 的总和。因此

$$N_T = A_T + N_{SF} + N_{BF} \tag{7.1}$$

2. 避雷器失效率

避雷器与大地之间形成一个瞬时通路，将输电线路上多余电荷泄入地下。最常见的避雷

器类型有开放式火花间隙、带火花间隙的碳化硅避雷器、无间隙的金属氧化物避雷器。第三种是目前最常用的避雷器,它由金属氧化物中的非线性镍铁铬合金组成,其中,最常见的是无火花间隙的氧化锌避雷器。

避雷器主要特点:

(1) 持续运行电压(U_C):指在避雷器末端之间长期持续运行的工频电压有效值。

(2) 额定电压:在暂态过电压下,避雷器末端正常运行状态下耐受的最大工频电压有效值。

(3) 残压(U_{res}):指放电电流通过避雷器时,其端子间出现的电压峰值。

(4) 放电额定电流:雷电冲击电流的峰值,用于对避雷器进行合理分类。

(5) 通流容量:避雷器可以自行冷却到正常运行温度的最大通流能力,而国家标准没有明确给出避雷器的通流容量。

在国际电力委员会(IEC)中,有"线路放电"这一术语,但由于基本信息不完全,因此,不同的设备制造商采用 kJ/kV(U_C)来衡量通流容量。依据国际电力委员会条款 8.5.5,热通流容量被定义为:当避雷器遭受两次脉冲电流时,其设备完好和耐热性稳定情况下的最大允许通流值。

New Words and Expressions

overhead lines	架空输电线路
high-voltage	高压
lightning protection	防雷保护;[电工学]避雷(装置)
electric power	[电]电能;功率
surge arrester	[电]避雷器;电涌放电器;冲击吸收器
overvoltage	n. [电]超电压,[电]过电压
lightning performance	防雷性能
distribution line	[电]配电线路
line outage	供电中断
power	n. 功率
	vt. 激励;动力
current rating	[电工学]额定电流
line tower	线路杆塔
bundled conductors	分裂导线
insulations	[电]绝缘
insulation level	绝缘水平;绝缘等级
footing resistance	[电]杆塔接地电阻
open loop	[电学]开环,开环回路
insulations flashovers	绝缘闪络
backflashover	n. 反向闪络,反击
operating voltage	[电]工作电压;运行电压
power frequency	工业频率

temporary overvoltage	暂时过电压
residual voltage	残余电压；[电工学]剩余电压
peak value	[电]峰值；最大值
discharge current	放电电流，泄漏电流
lightning current	雷电流；
impulse	n. [电子]脉冲

 Notes

1. Several studies have been conducted in an effort to either determine the minimum cost design of transmission lines, representing the total cost of the system as a function of the system design variables, or to explore the sensitivity of the required present worth of revenue, to several design variables, in order for design performance at minimum cost to be achieved.

目前，所有研究工作都是为获得使成本费用最小化的参数设计值，一方面是为了确定输电线路设计的最小经济成本，即与设计参数有关的数学函数来描述电力系统的总费用；另一方面，是为了探索当前经济成本的敏感因素。

2. Using a more sophisticated optimization algorithm, in contrast to this, based on direct computation, optimum values of those four parameters are calculated in order to minimize the defined performance indices.

与此相反，基于直接求解原理，提出了一种更加复杂的最优化算法，计算出四个参数的最优值，以使防雷性能评价指标最小化。

3. In IEC there exists the term line discharge class, but since this in not enough information, various manufacturers present thermal energy absorption capability in kJ/kV(U_c), defined as the maximum permissible energy that an arrester may be subjected to two impulses according to IEC clause8.5.5, without damage and without loss of thermal stability.

在国际电力委员会（IEC）中，有"线路放电"这一术语，但由于基本信息不完全，因此，不同的设备制造商采用 kJ/kV（U_c）来衡量通流容量。依据国际电力委员会条款 8.5.5，热通流容量被定义为：当避雷器遭受两次脉冲电流时，其设备完好和耐热性稳定情况下的最大允许通流值。

 **Exercises**

Ⅰ. Choose the best answer into the blank.

1. In this paper, the lightning protection of high-voltage transmission lines is faced as an _____problem.

    A. reduction   B. optimization   C. complex   D. extreme value

2. ____of power supply are very significant issues for energy utilities.

    A. Security and reliability   B. Security and sensitivity
    C. Rapidity and sensitivity   D. Rapidity and reliability

3. The proposed methodology only selects optimum values for the line _____ and the tower footing resistance.

  A. insulation level   B. voltage level   C. protection   D. flashover

4. It must be mentioned that the optimization methodology incorporates all the available protection means, i.e., _____ and surge arresters.

  A. bundled conductors   B. ground wires   C. shield wires   D. transmission line

5. A surge arrester presents a _____ path to earth, which removes the superfluous charge from the line.

  A. provisional   B. permanent   C. momentary   D. fixed

6. Continuous _____ is the rms value of power frequency voltage.

  A. rated voltage   B. residual voltage

  C. peak voltage   D. operating voltage

Ⅱ. Answer the following questions according to the text.

1. Why to choose Hellenic transmission lines of 150kV to validate the effectiveness of the proposed method?

2. Why program has developed to analyze lightning flashovers?

3. What should be related by suitable performance indices?

4. What is the purpose of using overhead ground wires and surge arresters?

5. How many types of arresters, and listed?

Ⅲ. Translate the following into Chinese.

Security and reliability of power supply are very significant issues for energy utilities. Supply interruptions are mainly related to faults in overhead high-voltage transmission lines, due to lightning strokes on them, producing dangerous overvoltages and damages on equipments. The correct and careful design constitutes the most important issue for transmission lines, since slight differences in the design parameter values can affect significantly their lightning performance, determining the operation and the performance of transmission lines for the rest of their life.

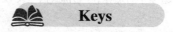

Keys

Ⅰ. Choose the best answer into the blank.

1. B   2. A   3. A   4. B   5. C   6. D

Ⅱ. Answer the following questions according to the text.

1. Because of their high failure rates during lightning thunderstorms.

2. Because when lightning strokes on them, they will produce dangerous overvoltage and damages on equipments.

3. Suitable performance indices are defined in order to relate the insulation level, the tower footing resistance, the energy absorption capability of the surge arrester and the surge arresters' installation interval cost values to the lightning failure costs.

4. The purpose of these protection methods is to reduce the insulations flashovers of the line that can cause power supply interruptions.

5. Four. The most common types of arresters are the open spark gaps, the SiC arresters with spark gaps and the metal oxide surge arresters without gaps. The last type, which is composed of non-linear resistors of metal oxide, mainly ZnO without spark gaps, is most commonly used today.

III. Translate the following into Chinese.

对于电网公司而言，安全性和可靠性是评价输电线路重要指标。雷击高压架空输电线路产生的电力设备过电压或机械性损坏，是导致电力系统供电中断的主要原因。输电线路最重要的一个环节是精准的参数设计，而细微的参数变化将显著地影响输电线路的防雷效果，并决定输电线路后续的运行状态和工作性能。

翻译技巧之状语从句（四）

关于表示让步的状语从句介绍如下。

1. 译成表示"让步"的分句

(1) Although he seems hearty and outgoing in public, Mr. Cooks is a withdrwn, introverted man.

虽然库克斯先生在公共场合中是热情而开朗的，但他却是一个孤僻、性格内向的人。

在上例中，两个分句的主语易位。

(2) At Paris, they were sheltered by the French members, because, although the Comet Line started in Belgium, it was very much a combined Belgium-French effort.

在巴黎，就由法国成员来掩护他们；因为，尽管彗星路线是在比利时开创的，但在很大程度上它是比利时人和法国人的共同事业。

(3) The threat of death does not depress him, even though he has become the No. 1villain to them.

即使他在他们心目中成了第一号坏蛋，死亡的威胁也不会使他消沉。

(4) While this is true of some, it is not true of all.

虽有一部分是如此，但不见得全部都是如此。

(5) While I grant his honesty, I suspect his memory.

虽然我对他的诚实没有异议，但我对他的记忆力却感到怀疑。

(6) I still think that you made a mistake while I admit what you say.

就算你说得对，我仍然认为你做错了。

"虽然""尽管""即使""就算"等是汉语中表示"让步"的常用关联词。汉语中表示"让步"的分句一般前置（但现在也逐渐出现后置现象），英语中则比较灵活。

2. 译成表"无条件"的条件分句

(1) Whatever combination of military and diplomatic action is taken, it is evident that he is having to tread an extremely delicate tight-rope.

不管他怎样同时采取军事和外教行动，他走的显然将不得不是一条极其脆弱的钢丝。

(2) No matter what misfortune befell him, he always squared his shoulders and said:

"Never mind, I'll work harder."

不管遭受到什么不幸事儿，他总是把胸一挺，说："没关系，我再加把劲儿。"

(3) Plugged into the intercommunication system, the man can now communicate with the rest of the crew no matter what noise is going on about him.

不管周围是多么喧闹，插头一接上机内通话系统，他就能和同机其余的人通话。

汉语里有一种复句，前一分句排除某一方面的一切条件，后一分句说出在任何条件下都会产生同样的结果，也就是说结果的产生没有什么条件限制。这样的复句里的前一分句，称之为"无条件"的条件分句。通常以"不论""不管""无论""管""虽"等作为关联词。

Section 2 Optimization of Hellenic Overhead High-voltage Transmission Lines Lightning Protection (2)

Momentary outages are a primary concern for some utilities trying to improve their power quality. Improving protection of overhead distribution circuits from lightning is one way in which some utilities can significantly reduce the number of momentary outages.

A computer program can help analyze various methods of improving the protection of distribution circuits without field tests. Many variables can be adjusted to measure their impact on the design:

(1) Conductor size and spacing;

(2) Insulation level;

(3) Grounding.

Power Technologies, Inc. has developed a program, the distribution Lightning Program (DLP), for analyzing lightning flashovers. Various lightning protection schemes can be simulated, including the use of shield wires and arresters. A good application of this program is to compare the use of arresters against the use of a shield wire. DLP was written to analyze distribution designs, but the calculation techniques and some of the results can be applied to transmission lines as well. This article provides background on lightning protection, broadly explains some of the traveling wave and lightning performance calculations, and provides examples and results.

1. Lightning Protection

Any lightning stroke to a distribution line will cause a flashover if no line protection is used. In addition, strokes that terminate near the line but do not actually hit it can induce voltages high enough to cause flashovers.

2. Power Quality

In most instances, flashovers result in a fault. The fault can be cleared by the substation breaker on the instantaneous setting of the relay, and the system will be back to normal following reclosure. This causes a momentary outage for the entire feeder that, in the past, may not have been a problem. However, modern consumer devices (such as digital clocks, computers, and VCRs) are disrupted by momentary outages, so many utilities are looking for ways to reduce these interruptions.

Lightning-caused flashovers from direct strokes can often be eliminated by some form of lightning protection using arresters and/or shield wires. These protection methods may be further improved by enhancements to insulation level and grounding. Lightning protect ion has the added benefit of reducing equipment damage and line burn-downs. Induced flashovers can also be reduced by improved design. Computer simulation can quickly tell how much improvement can be gained by these methods.

3. Insulation Level

The voltage level at which flashovers will occur on distribution structures is the basic impulse insulation level (BIL). Most distribution pole structures have a BIL from 150 to 400 kilovolts. BIL is important, because it determines whether the line flashes over due to voltages induced by a nearby lightning stroke. Accurate BIL measurements for structures are obtained by testing the structure with a surge generator, but estimates of BIL can also be made using insulator catalog values. The total BIL of a structure is not the same as the arithmetic sum of the BILs of individual components. Consider an example of a line where the lowest flashover path is across the insulator, over a length of wood, and across a fiberglass pin. The BIL is not the sum of the insulator BIL plus the BIL of the wood path plus the BIL of the pin. The structure BIL is smaller (in some cases much smaller) than the sum of the component BILs. There are many ways of estimating the total BIL, but they are beyond the scope of this article.

The use of concrete and steel structures on distribution lines is increasing, which could significantly reduce the structure's BIL. Also, metal crossarms and metal hardware are being used on wood structures. If such hardware is grounded, BIL will be reduced. The total BIL must be supplied by the insulator, and higher BIL insulators should be used to compensate for the loss of wood insulation. Obviously, trade-offs must be made between lightning performance and other considerations such as mechanical design or economics. The important point is to realize that trade-offs exist. The designer should be aware of the negative effects that metal hardware can have on lightning performance and try to minimize those effects.

Many types of equipment confuse the BIL issue. Guy wires, used to help hold poles upright, are generally attached as high on the pole as possible. These guy wires are effectively a grounding point if they do not contain insulating members; and if they are attached high on the pole, the BIL of the configuration will be reduced. The neutral wire height also affects BIL. On any given line, the neutral wire height may vary depending on equipment connected. On wood poles, the closer the neutral wire is to the phase wires, the lower the BIL. Many of the newer designs have lower BIL than older designs because of tighter phase spacing.

4. Line Protection

When lightning hits a line, it injects a current into the line, which splits into two with half of the current traveling in each direction at almost the speed of light. The surge voltage on the line for a direct stroke to the line can be estimated by using the surge impedance of the line, $V = ZI_S/2$, where V is the conductor voltage to ground, Z is the conductor surge impedance, and I_S is the stroke current. The surge impedance is a function of the inductance and capacitance of the line and the diameter of the corona envelope. Most overhead configurations have a surge impedance between 400 and 500 ohms. For an average lightning stroke current of 30 kiloamperes, the voltage on the stricken phase can reach over 6,000 kilovolts (400 ohms times 15,000 amperes), which is much higher than the BIL of distribution lines. Therefore, unless some sort of line protection is used, flashovers will occur due to direct strikes.

There are two types of line protection schemes typically associated with mitigating lightning-caused faults:

(1) Arresters;

(2) Shield wires.

Distribution arresters are primarily used to protect equipment such as transformers, but they can also be used to provide line protection. Liberal use of surge arresters can greatly enhance the protection of a distribution system provided they are properly applied. The mechanics of wave propagation with arresters present should be clearly understood, otherwise a small improvement may be bought at great cost.

Shield wires are used extensively to protect transmission lines, and they can also be used to protect distribution lines. Shield wires can provide improved protection if pole ground resistances are low and shield wire to phase spacings are adequate. Shield wires are placed so that virtually every lightning stroke will hit the shield wire. When lightning hits the shield wire, current will flow to ground through the ground impedance rather than entering the phase conductors.

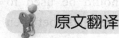
原文翻译

第二节　希腊高压架空输电线路防雷保护的最优化（2）

对于电力公司而言，暂时性供电中断是提高电能质量的最主要问题。优化架空配电线路的防雷保护设计，是很多电力公司减少暂时性供电中断的有效措施。

相比于现场试验，计算机程序可以帮助分析不同方法对提高配电线路防护效果的影响。在线路防雷设计过程中，参数可以自适应于不同的影响测量得到：

（1）导线规格和安装间隔；

（2）绝缘水平；

（3）是否接地。

电力技术公司已经研发了一款用于分析雷电防护的配电线路防雷软件（DLP）。该软件可以模拟包括避雷线和避雷器在内的多种防雷保护措施。这款软件的优点在于比较了避雷器与避雷线的防护效果。虽然这款软件是为了分析配电线路而设计，但其求解算法和某些研究结果同样适用于输电线路。本节介绍了防雷保护的背景知识，广泛地阐述了行波和雷电防护性能的计算方法，并给出了相应的算例和求解结果。

1. 防雷保护

如果线路上没有防雷保护措施，那么任何雷击配电线路将会遭受绝缘闪络。此外，当雷电击于线路末端附近而不是线路时，也会使配电线路电压升高至绝缘闪络。

2. 电能质量

在多数情况下，闪络会导致线路故障，变电站继电保护装置上的断路器将会跳闸以切除故障，在线路重合闸重新合闸后，系统恢复正常运行。在过去，这种情况将导致整条线路的暂时性断电，但这种情况在当时并未引起足够重视。然而，现代的暂时性断电将损坏用电设备（例如数字时钟、计算机和录像机），因此，很多电力公司正在致力于探索多种用于减少设备损坏的方法。

某些形式的防雷措施，例如避雷器或避雷线，可以消除由直击雷引起的绝缘闪络。随着绝缘水平和接地方式的改进，还能进一步地提升这些防雷措施的效果。雷电防护有助于减少设备损坏和线路烧毁引起的附加费用，而改进的线路设计同样可以减少诱导性绝缘闪络。同时，计算机仿真也可以快速地得到它们对绝缘闪络的改善效果。

3. 绝缘水平

配电线路上发生绝缘闪络的电压水平，被称为基准冲击绝缘水平（BIL），绝大多数配电线路的基准冲击绝缘水平在150kV到400kV之间。基准冲击绝缘水平是一个非常重要的参数，因为它决定了线路附近产生的感应雷击电压是否会导致绝缘闪络。精确的线路基准冲击绝缘水平可以通过脉冲发生器测试得到，同时，也可以通过绝缘子参数附表查询它的近似值。线路的基准冲击绝缘水平，并不是数学上的各独立部分基准冲击绝缘水平总和。例如，一条最短的绝缘闪络路径是同时贯穿绝缘子、一定长度的木杆和玻璃纤维制成的销子，而线路的基准冲击绝缘水平并不是绝缘子、木杆和销子的基准冲击绝缘水平之和。事实上，该值要小于（某些情况下更小）三者之和。当前阶段，有很多方法可以用来估算线路基准冲击绝缘水平数值，然而已经超出了本文的研究范畴。

在配电线路上，钢筋混凝土结构的使用比例正在逐渐增加，这会显著地降低线路的基准冲击绝缘水平。同时，在木质结构上，越来越广泛地应用金属叉梁和金属器件，如果这些金属器件接地，那么也会降低基准冲击绝缘水平。总的基准冲击绝缘水平必须由绝缘子计算得到，且应该应用更大数值的基准冲击绝缘水平来补偿因使用木质绝缘材料造成的基准冲击绝缘水平减少。显然，设计人员必须权衡防雷性能与其他若干因素，例如机械性能或经济性。关键之处在于，设计人员已经注意到权衡关系的存在，因此，为了降低金属器件对防雷性能的干扰，就应该使这种影响最小化。

不同种类的电气设备将会影响线路的基准冲击绝缘水平。拉线一般用于使电杆垂直，且固定点通常位于电杆的较高位置。如果不包含拉线的绝缘部分，那么拉线是一种有效的接地体；如果拉线固定点位于杆的高处，那将会降低线路的基准冲击绝缘水平。同时，零线的对地高度也会影响线路的基准冲击绝缘水平。对于任何给定线路，零线对地高度会随着各部件之间的连接关系而发生变化。在木质结构电杆上，相线与零线之间的几何距离越近，则基准冲击绝缘水平越低。由于相间结构布置更紧凑，很多新设计线路比旧线路基准冲击绝缘水平更低。

4. 线路保护

当雷电击中线路，会在线路上注入雷电流，且电流一分为二，以接近光的速度向着线路两侧进行独立传播。直击雷在线路上产生的冲击电压，可以采用线路的波阻抗进行估算，$V=ZI_s/2$，其中，V 是导线的对地电压，Z 是导线的波阻抗，I_s 为冲击电流，波阻抗是线路的电感、电容以及电晕包络面直径的数学函数。大多数架空线路的波阻抗在 400~500Ω 范围内。一般地，30kA 的平均雷电流在直击的相导线上产生的电压超过 6000kV（400Ω×15 000A），其数值远大于配电线路的基准冲击绝缘水平。因此，除非在线路上应用一些相应的防护措施，否则会直接导致绝缘闪络。

目前，有两种典型的以减少雷击故障的线路保护措施：

（1）避雷器；

（2）避雷线。

配电避雷器是电气设备的最主要保护方式，例如保护变压器，但是同时，它也可以为线路提供保护。合理正确的应用避雷器，将会显著地提高配电系统的运行能力。此外，设计人员需要充分了解现代避雷器的波传播原理，否则参数的微小变化将会产生巨额的经济成本。

避雷线被广泛地用于输电线路保护，同样也可以用于配电线路保护。如果杆塔接地电阻数值较小和足够的导地线相间距离，那可以充分地发挥避雷线的保护功能。安装避雷线，是为了让雷电每次都直接击中避雷线。当雷击中避雷线时，雷电流不是流向导线而是通过接地体直接泄入大地。

New Words and Expressions

power technologies	电力技术；电源技术；功率技术
shield wire	避雷线
traveling wave	［物理学］行（进）波，推进波
line protection	输电线路保护；输电线路保护装置
neutral wire	零线；中性线
phase wire	相线；三相三线制
phase spacing	相间距离
direct stroke	［电工学］直接雷
surge impedance	波阻抗；特性阻抗；［电］浪涌阻抗
inductance	n. 电感；感应系数；自感应
wave propagation	波传播
ground resistance	接地电阻
phase conductor	［电工学］相导线

Notes

1. The voltage level at which flashovers will occur on distribution structures is the basic

impulse insulation level (BIL).

配电线路上发生绝缘闪络的电压水平，被称为基准冲击绝缘水平（BIL）。

2. When lightning hits a line, it injects a current into the line, which splits into two with half of the current traveling in each direction at almost the speed of light.

当雷电击中线路，会在线路上注入雷电流，且电流一分为二，以接近光的速度向着线路两侧进行独立传播。

Exercises

Ⅰ. Choose the best answer into the blank.

1. Any lightning stroke to a distribution line will cause a flashover if no____ is used.
   A. line protection B. shield wire C. arrester D. insulator

2. The fault can be cleared by the____ on the instantaneous setting of the relay.
   A. line protection B. shield wire
   C. ground resistance D. substation breaker

3. Accurate BIL measurements for structures are obtained by testing the structure with a ____.
   A. relay B. surge generator C. neutral wire D. transformers

4. The designer should be aware of the____ effects that metal hardware can have on lightning performance.
   A. far-reaching B. negative C. positive D. side

5. The surge impedance is a function of the ____and ____of the line.
   A. resistance,inductance B. inductance,capacitance
   C. conductance,capacitance D. resistance,conductance

6. Shield wires can provide improved protection if pole ground resistances are ____ and shield wire to phase spacings are adequate.
   A. low B. high C. small D. huge

Ⅱ. Answer the following questions according to the text.

1. Which variables can be adjusted to measure their impact on the design?
2. Which program has developed to analyze lightning flashovers?
3. Why are many utilities looking for ways to reduce these interruptions?
4. What is surge impedance?
5. What is the operating principle of Shield wire?

Ⅲ. Translate the following into Chinese.

Lightning-caused flashovers from direct strokes can often be eliminated by some form of lightning protection using arresters and/or shield wires. These protection methods may be further improved by enhancements to insulation level and grounding. Lightning protect ion has the added benefit of reducing equipment damage and line burn-downs. Induced flashovers can also be reduced by improved design. Computer simulation can quickly tell how much improvement can be gained by these methods.

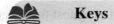
Keys

Ⅰ. Choose the best answer into the blank.
1. A   2. D.   3. B   4. B   5. B   6. A

Ⅱ. Answer the following questions according to the text.

1. Conductor size and spacing , insulation level ,grounding.

2. The distribution Lightning Program .

3. Because modern consumer devices (such as digital clocks , computers and VCRs) are disrupted by momentary outages.

4. The surge impedance is a function of the inductance and capacitance of the line and the diameter of the corona envelope.

5. Shield wires are placed so that virtually every lightning stroke will hit the shield wire. When lightning hits the shield wire, current will flow to ground through the ground impedance rather than entering the phase conductors.

Ⅲ. Translate the following into Chinese.

某些形式的防雷措施，例如避雷器和/或避雷线，可以消除由直击雷引起的雷电闪络。随着绝缘水平和接地方式的改进，还能进一步地提升这些防雷措施的效果。雷电防护有助于减少设备损坏和线路烧毁引起的附加费用，而改进的线路设计同样可以减少诱导性绝缘闪络。同时，计算机仿真也可以快速地得到这些防护措施对绝缘闪络的改善效果。

翻译技巧之状语从句（五）

关于表示目的的状语从句介绍如下。

1. 译成表示"目的"的前置分句

(1) They stepped into a helicopter and flew high in the sky in order that they might have a bird's-eye view of the city.

为了对这个城市做以鸟瞰，他们跨进了直升机，凌空飞行。

(2) He pushed open the door gently and stole out of the room for fear that he should awake her.

为了不惊醒她，他轻轻推开门，悄悄地溜了出去。

(3) We should start early so that we might get there before noon.

为了正午以前赶到那里，我们很早就动身了。

汉语里表"目的"的分句所常用的关联词有"为了""省（免）得""以免""以便""生怕"等，"为了"往往用于前置分句，"省（免）得""以免""以便""使得""生怕"等一般用于后置分句。

2. 译成表示"目的"的后置分句

(1) The murderer ran away as fast as he could, so that he might not be caught red-handed.

凶手尽快地跑开，以免被人当场抓住。

(2) Brackett groaned aloud: "You came from Kansas City in two week so that I could give you a job?"

布雷克特唉声叹气地说："你从堪萨斯城走了两个星期到这里，就是要我给你找个工作吗？"

(3) Wouldn't it be better if we were somehow to organize an escape route so that these men could get back to Britain into the war again?

要是我们有点什么办法来安排一条可供安全逃脱的路线，使这些人能回到英国重新投入战争，这不更好吗？

Section 3　Modeling of Power Transmission Lines for Lightning Back Flashover Analysis

Performance of power transmission lines has a great impact on reliability of a particular power supply system of a country. Unreliable power transmission lines can even lead to total power failures resulting with great financial losses. The lightning back flashover effects are recognized as one of the major causes of transmission line outages, especially in tropical countries with frequent lightning. Several types of solutions are presently available to address the issue of lightning back flashovers. However, the modern concept of transmission line mounted surge arresters is of great popularity due to its excellent performance, ease of installation and the low cost compared to the other solutions.

This paper describes a case study carried out on one of the critical hill-country 220kV power transmission lines of the Sri Lankan transmission network, having several past records of lightning back flashover related outages leading to total system failures. The study described in this paper mainly focuses on the way of analyzing the back flashover event by transient modeling and subsequent simulation of the selected transmission line in an electromagnetic transient computer program. The study uses the Power System CAD (PSCAD) software program as the software tool for the purpose of modeling and simulation.

Simulations are carried out with and without Transmission Line Arrester (TLA) to evaluate TLA's impact on back flashover. The results of the simulations show that the installation of 02nos. of TLA at top phases of each selected towers improve the overall performance.

Protection of transmission lines against lightning back flashovers is considered as one of the major research areas in modern power sector, as the back flashovers are recognized as one of the major reason of line failures. Back flashover events occur when the lightning strikes on either tower or shield wires. These strikes produce waves of currents and voltages travelling on the shield wires called travelling waves and reflections occurs at every points where impedance discontinuities.

Accordingly surge voltages can be developed across line insulators exceeding the Critical Flashover Voltage where flashovers occur from tower to line called back flashovers.

The case study described in this paper is based on one of the most critical 220kV double circuit power transmission line which connects one of the major hydro generation complexes of Sri Lanka to the main load centre at Colombo. According to the past performance records of this transmission line, it has been noticed that the failure of this transmission line has great influence towards a total failure of the system.

Out of those, most of the line outages were due to the effect of lightning, since most of them were recorded in the months, April-June and October-November where the lightning is frequent. It has been found in a previous study, that there are two line sections of the selected transmission line having higher probability of insulator damages and flashover marks according to the breakdown and maintenance records. Those line sections have been selected for the study described in this paper.

The transmission line selected in this study is about 70.5km in length and having 206nos. of double circuit steel lattice towers with 2nos. galvanized steel overhead ground wires providing protection against lightning.

1. Transient Model

The methodology of analyzing the effects of lightning back flashovers on the selected 220kV Biyagama-Kotmale power transmission line consists of fast front transient modeling and simulation using electromagnetic transient type software. The power transmission line and the back flashover event were modeled by frequency dependent fast front transient models due to the nature of higher frequency dependency of lightning strokes.

The basic fast front transient transmission line model which was implemented in the PSCAD (Power Systems Computer Aided Design) is shown in the fig.7.1. The complete line model consists of several sub models representing the following transmission line elements.

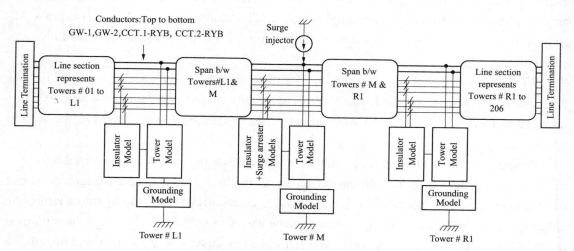

Fig.7.1 Complete transmission line model used for the analysis

(1) "Transmission line sections" including towers up to the line end terminations (ex: line section with towers from tower no.01 to L1 as shown in the fig.7.1).

(2) "Transmission line spans" between consecutive towers under study (ex: span between tower no. L1 to M as shown in the fig.7.1).

(3) Transmission tower.

(4) Tower grounding resistance.

(5) Line insulator string with back flashover model.

(6) Line end terminations.

(7) Line Surge Arrester.

(8) Lightning surge generator model.

(9) Power frequency phase voltage generator model.

As shown in fig.7.1, the complete line model consists of three towers named by tower number M, L1 and R1 which represents a typical tower and two adjacent towers at the left and right sides of it respectively. The two line spans between these three towers were represented by the "line span models" whereas the rest of the line sections at each side up to the end terminations were represented by "line section models". Six number of inter connecting lines were used to connect each modules while representing the ground and phase conductors from top to bottom sequence. All three tower models are connected to the ground wire-1 (GW-1) and ground wire-2 (GW-2) whereas the connections to the phase conductors are made through the insulator models.

2. Simulation Criteria

Three (03) numbers of variables those directly affecting the back flashovers were varied in the series of the simulation runs. The variables are Magnitude of the lightning surge current, Phase angle of the power frequency phase voltage and the grounding resistance of Tower-M.

Simulation of complete line model is carried out in two major steps. At first step, the model is simulated without arresters for selected three (03) critical tower grounding resistance values 9, 17Ω and 62Ω for both 8×20μs and 1.2×50μs surge waveforms.

In the second step the model is simulated with surge arresters with two different arrester configurations for both 8×20μs and 1.2×50μs surge waveforms. The simulations were carried out for above two arrester configurations with 62Ω tower grounding resistance which is the worst case out of the selected three values.

For all simulations the lightning surge current was injected on top of the Tower M.

第三节 用于雷电反击分析的输电线模型

输电线路的性能对一个国家特定的电力供应系统的可靠性有着重要影响。不可靠的输电线路甚至能导致整个电力系统的故障，从而造成严重的经济损失。雷电反击效应被认为是造成输电线路停运的主要影响之一，特别是对有频繁雷电的热带国家。针对雷电反击的问题，目前提出了几种类型的解决方案。然而，在输电线路上安装避雷器的现代概念更受欢迎，因为避雷器具有出色的防雷性能，并且安装简便，同时相比于其他解决方法成本更为低廉。

本文介绍了斯里兰卡输电网中一条临界丘陵区 220kV 输电线路的案例研究。该线路在过去发生了几次雷电反击停运事故，造成了全系统发生故障。该研究的重点在于利用瞬态模型分析雷电反击事故的方法，以及随后通过电磁暂态计算机程序对所选输电线路进行反击模拟。在该研究中，利用电力系统 CAD（PSCAD）软件程序作为软件工具，来进行建模与仿真。

分别在有、无输电线路避雷器（TLA）的情况下进行仿真，以此评估避雷器对雷电反击的影响。仿真结果表明，在选定杆塔顶相安装 2 号避雷器可以提升整体性能。

由于雷电反击是造成输电线路故障的主要原因之一，因此输电线路雷电反击保护被认为是电力行业的主要研究领域之一。雷电反击事故发生在雷击杆塔或屏蔽线时，这些雷击将会在屏蔽线上产生被称为行波的电流波和电压波，同时在每个波阻抗不连续的点产生反射波。

因此，当超过临界击穿电压时，从杆塔到线路的区域将会发生闪络（称为反击），因而冲击电压能够穿过线路绝缘子。

本文介绍的案例研究是基于最重要的 220kV 双回输电线路当中的一条，其将斯里兰卡一个主要的水力发电综合体与科伦坡主要的负载中心连接起来。根据这条输电线路过去的性能记录，可以发现它的故障对系统的整体故障有着重要影响。

线路停运的大多数原因归结于雷电的影响，因为大部分事故都是在雷电频发的月份，如六月到八月、十月到十一月被记录的。在以前的研究中发现，根据故障和维护记录，在所选输电线路中，有两条线段的绝缘子损坏与闪络概率较高。因此，在本文介绍的案例分析汇中选择了这些线段来进行研究。

本文所选的输电线路长度约为 70.5km，包含 206 基双回钢结构杆塔。为防止雷击事故，在杆塔架设了 2 条架空镀锌钢地线。

1. 瞬态模型

分析 220kV 比耶加默—柯特梅利输电线路雷电反击影响的方法，包括快速前沿瞬态建模和利用电磁暂态软件仿真。由于雷击对频率具有高依赖性，因此通过与频率相关的快速前沿瞬态模型来模拟输电线路和雷电反击事件。

在软件 PSCAD（一种电磁暂态仿真软件）中搭建的基本快速前沿瞬态模型如图 7.1 所示。该完整的线路模型包括一些代表如下输电线路元件的子模型。

（1）"输电线路部分"包括杆塔上端到线路终端的部分（不包括：如图 7.1 所示从 01 号到 L1 号杆塔之间的线段）。

（2）研究范围内的相邻杆塔之间的"输电线路挡距"（不包括：如图 7.1 所示 L1 号杆塔到 M 号杆塔之间的挡距）。

（3）输电杆塔。

（4）杆塔接地电阻。

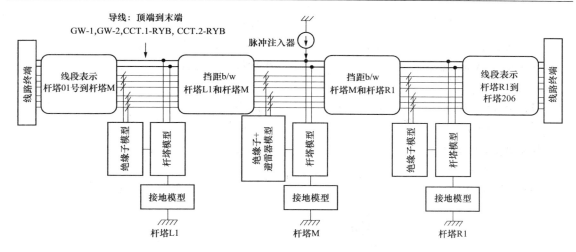

图 7.1　用于分析的完整输电线路模型

（5）带有反击模型的线路绝缘子串。
（6）线路终端。
（7）线路避雷器。
（8）雷击脉冲发生器模型。
（9）工频相电压发生器模型。

如图 7.1 所示，该完整的线路模型包括编号为 M、L1 和 R1 三个杆塔，它们分别代表一个典型杆塔，以及与该典型杆塔左右相邻的两个杆塔。这三基杆塔之间的两个线路挡距用"线路挡距模型"表示，而线段每侧上端到末端的其他部用"线段模型"表示。各模块之间用六组内部连接线相连，用以表示从上到下相导的接电线和相导线。所有的三基杆塔模型用避雷地线 1（GW-1）和避雷地线 2（GW-2）连接，而相导线之间通过绝缘子模型连接。

2. 仿真条件

在进行一系列仿真运行时，改变直接影响雷电反击的三个变量值。这三个变量分别是雷电冲击电流的幅值、工频相电压的相位角和 M 号杆塔的接地电阻。

运用以下两个主要步骤来进行完整线模型的仿真。第一步，分别在选定的三个 9、17、62Ω 临界杆塔接地电阻，以及 8×20μs 和 1.2×50μs 的冲击波形下，用不带避雷器的模型来进行仿真。

第二步，分别在 8×20μs 和 1.2×50μs 的冲击波形，以及 62Ω 杆塔接地电阻（在上述三个所选的杆塔接地电阻值中，该电阻产生的影响最坏）下，用带避雷器的模型进行仿真。

对于所有的仿真，雷电冲击电流均是从杆塔 M 的顶端注入的。

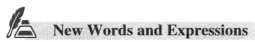 **New Words and Expressions**

power failure	停电；[电] 电源故障
electromagnetic	*adj.* 电磁的
electromagnetic transient	电磁暂态
phase	*n.* 相；阶段；[天] 位相

top phase	顶相，上相
power sector	电力部门；电力工业
travelling wave	行波；前行波
surge voltage	［电］冲击电压
flashover voltage	闪络电压；击穿电压
double circuit	双回路
hydro generation	水力发电厂
load centre	负载中心；负荷中心
lattice towers	［无线电］格架塔（用来支持天线）

Notes

1. However, the modern concept of transmission line mounted surge arresters is of great popularity due to its excellent performance, ease of installation and the low cost compared to the other solutions.

然而，在输电线路上安装避雷器的现代概念更受欢迎，因为避雷器具有出色的防雷性能，并且安装简便，同时相比于其他解决方法成本更为低廉。

2. These strikes produce waves of currents and voltages travelling on the shield wires called travelling waves and reflections occurs at every points where impedance discontinuities.

这些雷击将会在屏蔽线上产生被称为行波的电流波和电压波，同时在每个波阻抗不连续的点产生反射波。

Exercises

Ⅰ. Choose the best answer into the blank.

1. The modern concept of transmission line mounted ＿＿＿ is of great popularity.

　　A. surge arresters　　　　　　　B. ground wires
　　C. lightning rods　　　　　　　 D. lightning conductions

2. The results of the simulations show that the installation of 02nos. of TLA at ＿＿＿ of each selected towers improve the overall performance.

　　A. top phases　　B. milldle phases　　C. first phase　　D. last phase

3. These strikes produce waves of currents and voltages travelling on the＿＿＿ called travelling waves.

　　A. lines　　　　B. towers　　　　C. shield wires　　D. surge arresters

4. The case study described in this paper is based on one of the most critical 220kV ＿＿＿ power transmission line.

　　A .single circuit　　　　　　　B. double circuit
　　C. triple circuit　　　　　　　 D. single or double circuit

5. The complete line model consists of ＿＿＿ towers.

Chapter 7　Lightning Protection and Grounding　　151

  A. two　　　　　　B. three　　　　　　C. four　　　　　　D. five

  6. The simulations were carried out for above two arrester configurations with ____ tower grounding resistance which is the worst case out of the selected three values.

  A. 9Ω　　　　　　B. 17Ω　　　　　　C. 62Ω　　　　　　D. 9Ω and 17Ω

  II. Answer the following questions according to the text.

  1. What will happen if power transmission lines are unreliable?

  2. Which software program did the study in this paper choose for modeling and simulation?

  3. Were most of the line outages due to the effect of lightning?

  4. What were modeled by frequency dependent fast front transient models in this paper?

  5. How many numbers of variables those directly affecting the back flashovers were varied in the series of the simulation runs, and listed?

  III. Translate the following into Chinese.

  Protection of transmission lines against lightning back flashovers is considered as one of the major research areas in modern power sector, as the back flashovers are recognized as one of the major reason of line failures. Back flashover events occur when the lightning strikes on either tower or shield wires. These strikes produce waves of currents and voltages travelling on the shield wires called travelling waves and reflections occurs at every points where impedance discontinuities.

Keys

  I. Choose the best answer into the blank.

  1.A　　2.A　　3.C　　4.B　　5.B　　6.C

  II. Answer the following questions according to the text.

  1. Unreliable power transmission lines can even lead to total power failures resulting with great financial losses.

  2. PSCAD.

  3. Yes.

  4. The power transmission line and the back flashover event.

  5. Three. The variables are Magnitude of the lightning surge current, Phase angle of the power frequency phase voltage and the grounding resistance of Tower-M.

  III. Translate the following into Chinese.

  由于雷电反击是造成输电线路故障的主要原因之一，因此输电线路雷电反击保护被认为是电力行业的主要研究领域之一。雷电反击事故发生在雷击杆塔或屏蔽线时，这些雷击将会在屏蔽线上产生被称为行波的电流波和电压波，同时在每个波阻抗不连续的点产生反射波。

翻译技巧之长句（一）

  1. 英语长句

  英语长句一般指的是各种复杂句，复杂句里可能有多个从句，从句与从句之间的关系可能包孕、嵌套，也可能并列、平行。所以翻译长句，实际上重点主要放在对各种从句的翻译上。从功能来说，英语有三大复合句：①名词性从句，包括主语从句、宾语从句、表语从句和同位语从句；②形容词性从句，即平常所说的定语从句；③状语从句。

2. 英语长句的特点

一般说来，英语长句有如下几个特点：

（1）结构复杂，逻辑层次多；
（2）常需根据上下文做词义的引申；
（3）常需根据上下文对指代词的指代关系做出判断；
（4）并列成分多；
（5）修饰语多，特别是后置定语很长；
（6）习惯搭配和成语经常出现。

3. 英语长句的分析方法

（1）找出全句的主语、谓语和宾语，即句子的主干结构；
（2）找出句中所有的谓语结构、非谓语结构、介词短语和从句的引导词；
（3）分析从句和短语的功能，例如，是否为主语从句、宾语从句、表语从句或状语从句等，以及词、短语和从句之间的关系；
（4）分析句子中是否有固定词组或固定搭配、插入语等其他成分。

Chapter 8　Stability Problems of Power Grid

Section 1　Major Stability Problems of Long Distance and Large Capacity High Voltage AC/DC Transmission Systems and Interconnected Power Systems

In recent years, as long distance hybrid AC/DC power transmission systems and nationwide interconnected power system are developed, the stability of power systems has become more important in China. Now, the most challenging issues of China's power grid include transient stability of interconnected power systems, low frequency oscillation, voltage stability, and shortcircuit current level control.

1. Transient stability

Transient instability, which is triggered by large disturbances, is still one of the main stability issues of power systems. In recent years, with the increase of long distance AC/DC transmission lines with large transmission capacity and the emergence of the nationwide interconnected power system, new transient stability issues have arisen in China's power systems.

Major transient stability issues of China's power gird are as follows.

(1) A few 500kV AC lines are used to interconnect large regional power grids, so the interconnections between these large power grids are very weak. As a result, the stability levels of some transmission corridors within the large regional power grids have declined after interconnection.

(2) There are many irrational structures within some power grids. Electromagnetic loop networks still exist in the 500kV and 220kV power grids, significantly reducing transmission capacity.

(3) There are stability problems in the hybrid AC/DC power grids. The South China Power Grid is just a hybrid AC/DC power grid. If two poles of a DC line are blocked and AC lines in parallel with the DC line are then overloaded, large-scale power transfer may cause power system to lose synchronization.

China has paid much attention to the stability of power systems. The standard for Security and Stability of Power Systems was passed at the Dalian Power System Sability Meeting in 1981, and amended in 2001. The standard requests that three defense lines for security protection of power grids must be built. Currently, since China's ultra high voltage power grid is still relatively weak, rational arrangement of relay and automatic security devices are top priorities set by each large regional grid. Additionally, generator and load tripping are widely adopted to ensure safe operations of the power systems. With these efforts, China's power system will be kept stable even if a very serious disturbance takes place.

2. Voltage stability

Maintaining node voltages at predetermined levels will ensure not only the quality of the power supply, but also the stability of the power system operations. A device failure caused by abnormal voltage, voltage instability, and voltage collapse can trigger power system blackouts. For example, the nationwide blackout in France in December 1978, the Quebec Power System blackout in Canada in December 1982, the blackout in Sweden in November 1983, and the blackout in Tokyo Power System in July 1987. To avoid the blackout caused by voltage collapse has become a very important issue in system design and operation. In China, because of the rapid development of modern power systems, some receiving power systems with highly dense loads have emerged, such as Guangdong Power Grid in southern China, Shanghai Power Grid in eastern China, Beijing-Tianjin-Tangshan Power Grid in northern China.

Hierarchical and multi-area control is used to control voltage and reactive power. In general, voltage control can be categorized into three levels. The first level is located at power plants or customer sides, with functions, such as regulation of reactive power of generators and output of SVCs, and fast switch of capacitances and reactors. These controls usually take seconds to respond at this level. The second level is usually located at pivotal sites, and the response time is typically several minutes. This type of control is used to coordinate primary control on the spot, for example, adjusting the voltage regulation set points of generators and SVCs, and switching capacitances and reactors and set reference voltages of buses. The third level is preventive control, and its response time spans tens of minutes. The objectives of this type of control include identifying potential voltage instability in order to take necessary measures, coordinating the second level control systems, and optimizing voltage and reactive power. Additionally, the static voltage stability analysis is also considered in the third level of control. Several dispatching bureaus in China, such as the North China Power Grid Control Center, Jiangsu Power Grid Control Center, Fujian Power Grid Control Center, have installed automatic voltage control software (AVC).

3. Low frequency oscillation

A low frequency oscillation is caused by negative damping torques that typically take place in power systems with generators connected with main power grids using long distance lines. Currently, weak or even negative damping torques exist in China's major interconnected power grids, and low frequency oscillation took place in many power grids, such as the South China Power Grid in 1994, Ertan Power Transmission System in 1998 and 2000, Guangdong-Hongkong Interconnected Power Grid in 2003. In the future, more large capacity generators and fast exciters will be put in use in major interconnected power grids. As a result, a weak damping will happen and possibly escalate.

The power system stabilizer (PSS) can be applied to enhance system damping and suppress low or ultra low frequency oscillation. In recent years, the State Grid Corporation of China has made many efforts to study how to use PSS to maintain the dynamic stability of interconnected power systems. As a result, many PSSs have been installed in power grids, and the dynamic stability level has been greatly improved.

4. High short circuit current level

With rapid load increase and the interconnection of large regional power grids, short-circuit current level increased over the years. In recent years, very high short-circuit current level has become a major issue that must be addressed in the stage of planning and operation of power systems. For example, the shortcircuit current of pivotal stations in the delta area of the Yangzi River in the East China Power Grid approached or even exceeded the switching capacity of breakers. In 2010, the short-circuit current of the Doushan, Huangdu, Wunan, Wangdian, Lanting substations will exceed switching capacities of their breakers, and the short-circuit current of Wunan and Wangdian substations will even exceed 63kA. High short-circuit current has become one of the most serious issues of the East China Power Grid.

Short-circuit current level is directly related to the structure, density, and intensity of the power grid, as well as commitment mode of generators in the power systems. To control the level of short-circuit current, the structure of power sources and power grids must be taken into account to achieve optimization.

5. Improving thermal stability limit of transmission lines

With the rapid economic growth, the demand for electricity has increased dramatically in recent years, especially in the south Jiangsu Province, Shanghai and Zhejiang provinces. In these areas, the power transmission capabilities of some transmission lines are seriously limited by lack of thermal stability. For example, after the completion of the DC line from the Three Gorges to Changzhou, the required power transmission capabilities of the 500kV line from Wunan to Doushan and the double circuit lines from Doushan to Shipai have already exceeded the capacity limited by the levels of thermal stability. As a result, it is urgent for the East China Power Grid to enhance the transmission capability of 500kV transmission lines. Currently, there are a set of new rules and techniques to be adopted by the East China Grid Company Limited to enhance the transmission capability of 500kV transmission lines.

(1) At the end of 2005, the East China Grid Company Limited increased the highest operation temperature of 500kV transmission lines from 70℃ to 80℃; consequently, the transmission capability of parts of these 500kV transmission lines increased from 1900MW to 2700MW.

(2) Using a short-time overloading capability of 500kV transformers to increase transmission capability.With thismethod, the short-time overloading capabilities of most 500kV major transformers have increased 1.3～1.5 times, and transmission capabilities of most 500kV major transformers increased by 7%～15%.

(3) Use real-time monitoring of current, transmission line temperature, external temperature, wind direction and velocity, and illumination density.

原文翻译

第一节 长距离、大容量高压交/直流输电系统与互联电力系统中的主要稳定问题

近年来，随着远距离交直流混合输电系统和全国互联电力系统的发展，电力系统的稳定性在中国已变得十分重要。现在，中国电网面临的最具挑战性的问题包括互联电力系统的暂态稳定性、低频振荡、电压稳定和短路电流水平的控制。

1. 暂态稳定

由大扰动触发的暂态不稳定性仍是电力系统稳定性的一个主要问题。近年来，随着大容量长距离交/直流输电线路的增加和全国互联电力系统的出现，中国电力系统暂态稳定性出现了新的问题。

中国电网主要的暂态稳定性如下。

（1）大区域的电网是由一些500kV交流线路连在一起的，因此这些大电网之间的相互关系非常弱。这导致一些输电线路的稳定性在大区域电网互联后已经有所下降。

（2）很多电网内部存在一些不合理的结构。在500kV和220kV电网里，电磁环网仍然存在，这大大降低了线路传输能力。

（3）交直流混合电网存在稳定性问题。南方电网是一个交直流混合电网。如果直流线路的两极都被封锁了，交流线路与直流线路并行，然后超负荷，大规模的输电可能导致电力系统失去同步。

中国高度重视电力系统的稳定性。电力系统的安全与稳定的标准是在1981年大连电力系统稳定会议上通过的，并于2001年修订。标准要求必须建立三道防线来对电网进行安全保护。由于当前中国特高压电网仍然比较薄弱，因此合理地安排继电保护和自动安全装置是各大区域电网优先考虑的问题。此外，发电机和负荷的切换也被广泛地采用以确保电力系统安全运行。通过这些努力，即使发生非常严重的干扰，中国的电力系统也将保持稳定。

2. 电压稳定

节点电压维持在预定水平不仅将确保供电质量，而且还可以保证电力系统运行的稳定性。由电压异常、电压不稳定和电压崩溃引起的设备故障可能导致电力系统发生停电事故。例如，1978年12月发生在法国的全国大面积停电事故，1982年12月发生在加拿大魁北克的电力系统停电事故，1983年11月在瑞典的停电事故，以及1987年7月在东京电力系统停电事故。如何避免电压崩溃导致的停电已成为系统设计和运行中的一个非常重要的问题。在中国，由于现代电力系统的迅速发展，出现了一些负荷高度密集的受电系统，如在中国南部的广东电网、中国东部的上海电网以及中国北部的北京—天津—唐山电网。

分层多领域控制可用来控制电压和无功功率。一般而言，电压控制可分为三个级别。第一级位于电厂或客户方面，具有类似发电机无功功率调节和静止无功补偿装置以及开关电容、电抗器的功能。这些控件在这一级响应通常需要数秒。第二级通常位于关键地点，响应时间通常是几分钟。这种类型的控制是用来协调现场主要控制，例如，调整发电机和静止无功补偿装置的电压调节点，以及调整开关电容、电抗器和设置总线的参考电压。第三级是预防控制，其响应时间长达十分钟。这种类型控制的目标主要是确定潜在电压不稳定性，以便采取必要措施，协调二级控制系统和优化电压及无功功率。此外，静态电压稳定分析也被认为是第三控制等级。在中国的几个调度部门，如华北电网控制中心、江苏电网控制中心、福建电网控制中心，都已经安装了自动电压控制软件（AVC）。

3．低频振荡

低频振动是由负阻尼力矩造成，它通常发生在发电机与主电网采用长距离线路连接的电力系统中。目前，弱甚至负阻尼力矩问题通常存在于中国主要的互联电网，低频振荡在许多电网都有发生，如 1994 年南方电网，1998 和 2000 年二滩输电系统，2003 年广州—香港互联电网。在未来，更大容量的发电机和快速励磁机将在大型互联电网投入使用。因此，弱阻尼会发生，并有可能加剧。

电力系统稳定器（PSS）可以用来提高系统的阻尼，抑制低或超低频率振动。近年来，中国国家电网公司已做出许多努力来研究如何利用电力系统稳定器保持互联电力系统的动态稳定性。因此，在电网中已经安装了许多电力系统稳定器，使得动态稳定水平有了很大改善。

4．短路电流水平较高

这些年来，随着负荷的增加和大区域电网的互联，短路电流水平也在逐步提高。近年来，非常高的短路电流水平在电力系统规划和运行阶段已成为一个必须解决的重大问题。例如，在长江三角洲地区或华东电网，关键点的短路电流已经接近或超过断路器的开关能力。2010 年，斗山、黄渡、乌南、王店、兰亭变电站的短路电流将超过断路器的开断容量，梧南和王店变电站的短路电流将超过 63kA。高短路电流已经成为华东电网最严重的问题之一。

短路电流水平直接关系到电网的结构、密度和供电可靠性，以及电力系统中发电机的运行模式。为了控制短路电流水平，必须优化电源和电网的结构。

5．提高输电线路的热稳定极限

近年来随着经济的快速增长，我国的电力需求急剧增加，特别是江苏、上海和浙江省。在这些地区，一些输电线路的输电能力严重受限于较低的热稳定性。例如，在三峡至常州的直流线路完工后，梧南至斗山 500kV 输电线路和斗山至石牌 500kV 双回输电线路的输电容量已超过其热稳定性限值。因此，华东电网 500kV 输电线路的输电容量迫切需要提高。目前，华东电网有限公司提出了一系列新的规范和技术来加强 500kV 输电线路的输电容量。

（1）截至 2005 年年底，华东电网有限公司将 500kV 输电线路最高运行温度从 70℃ 提升到了 80℃；因此，部分 500kV 输电线路传输容量从 1900MW 增加到 2700MW。

（2）利用 500kV 变压器的短时过负荷能力可以提高输电容量。使用该方法后，500kV 主要变压器的短时过负荷容量提升了 1.3～1.5 倍，其输电容量增加了 7%～15%。

（3）采用电流、输电线路温度、环境温度、风向、风速以及光照密度的实时监控。

New Words and Expressions

electromagnetic loop networks	电磁回路网络
hierarchical	*adj.* 分层的，等级体系的
blackou	*n.* 灯火管制，灯火熄灭
frequency oscillation	频率振荡

damp	n. 潮湿，湿气
	adj. 潮湿的
	vt. 使潮湿；使阻尼；使沮丧，抑制
	vi. 减幅，阻尼；变潮湿
short-circuit	vt. 使短路
	vi. 发生短路
automatic	n. 自动机械；自动手枪
	adj. 自动的；无意识的；必然的
intensity	n. 强度；强烈；[电子] 亮度；紧张
transient stability	[物] 暂态稳定性
low frequency oscillation	低频振荡
short circuit	短路
disturbance	n. 干扰
in parallel with	并联
overload	n. 过负荷
	vt. 过负荷
synchronization	n. 同步
relay	n. [电] 继电器
generator	n. [电] 发电机
node	n. 节点
blackout	停电
reactive power	[电] 无功功率
capacitance	n. [电] 电容；电流容量
reactor	n. 电抗器
negative damping	负阻尼
breaker	n. [电] 断路器
node voltage	[电] 节点电压

 **Notes**

1. In recent years, as long distance hybrid AC/DC power transmission systems and nationwide interconnected power system are developed, the stability of power systems has become more important in China.

近年来，随着远距离交直流混合输电系统和全国互联电力系统的发展，电力系统的稳定性在中国已变得十分重要。

2. Maintaining node voltages at predetermined levels will ensure not only the quality of the power supply, but also the stability of the power system operations.

节点电压维持在预定水平不仅将确保供电质量，而且还可以保证电力系统运行的稳定性。

3. A low frequency oscillation is caused by negative damping torques that typically take place

in power systems with generators connected with main power grids using long distance lines.

低频振动是由负阻尼力矩造成，它通常发生在发电机与主电网采用长距离线路连接的电力系统中。

4. Currently, there are a set of new rules and techniques to be adopted by the East China Grid Company Limited to enhance the transmission capability of 500kV transmission lines.

目前，华东电网有限公司提出了一系列新的规范和技术来加强500kV输电线路的输电容量。

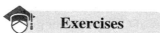

Exercises

I. Choose the best answer into the blank.

1. The interconnections between these large power grids are very ____.
   A. well   B. bad   C. strong   D. week

2. The standard requests that ____ defense lines for security protection of power grids must be built.
   A. one   B. two   C. three   D. four

3. ____ can be applied to enhance system damping and suppress low or ultra low frequency oscillation.
   A. SVC   B. PSS   C. AVC   D. AC/DC

4. The East China Grid Company Limited increased the highest operation temperature of 500kV transmission lines from ____ to ____.
   A. 40；50   B. 50；60   C. 60；70   D. 70；80

5. From ____ to ____ and the double circuit lines from Doushan to Shipai have already exceeded the capacity limited by the levels of thermal stability.
   A. Wunan;Doushan   B. Doushan;Huangdu
   C. Wunan;Wangdian   D. Huangdu;Changzhou

II. Answer the following questions according to the text.

1. What is the major transient stability issues of China's power gird?
2. What is hierarchical and multi-area control is used to control?
3. What dispatching bureaus have installed the automatic voltage controlsoftware?

III. Translate the following into Chinese.

With the rapid economic growth, the demand for electricity has increased dramatically in recent years, especially in the south Jiangsu Province, Shanghai and Zhejiang provinces. In these areas, the power transmission capabilities of some transmission lines are seriously limited by lack of thermal stability. For example, after the completion of the DC line from the Three Gorges to Changzhou, the required power transmission capabilities of the 500kV line from Wunan to Doushan and the double circuit lines from Doushan to Shipai have already exceeded the capacity limited by the levels of thermal stability. As a result, it is urgent for the East China Power Grid to enhance the transmission capability of 500kV transmission lines. Currently, there are a set of new rules and techniques to be adopted by the East China Grid Company Limited to enhance the transmission capability of 500kV transmission lines.

 **Keys**

I. Choose the best answer into the blank.
1. D   2. C   3. B   4. D   5. A

II. Answer the following questions according to the text.

1. Major transient stability issues of China's power gird are as follows.

(1) A few 500kV AC lines are used to interconnect large regional power grids, so the interconnections between these large power grids are very weak.

(2) There are many irrational structures within some power grids.

(3) There are stability problems in the hybrid AC/DC power grids.

2. voltage and reactive power.

3. The North China Power Grid Control Center, Jiangsu Power Grid Control Center, Fujian Power Grid Control Center.

III. Translate the following into Chinese.

近年来随着经济的快速增长，我国的电力需求急剧增加，特别是江苏、上海和浙江省。在这些地区，一些输电线路的输电能力严重受限于较低的热稳定性。例如，在三峡至常州的直流线路完工后，梧南至斗山500kV输电线路和斗山至石牌500kV双回输电线路的输电容量已超过其热稳定性限值。因此，华东电网500kV输电线路的输电容量迫切需要提高。目前，华东电网有限公司提出了一系列新的规范和技术来加强500kV输电线路的输电容量。

翻译技巧之长句（二）

关于长句翻译方法介绍如下：

In Africa I met a boy, who was crying as if his heart would break and said, when I spoke to him, that he was hungry because he had had no food for two days.

分析：

第一，拆分句子：这个长句可以拆分为四段：In Africa I met a boy/who was crying as if his heart would break/when I spoke to him, that he was hungry because/he had had no food for two days.

第二，句子的结构分析：

（1）主干结构是主语+过去式+宾语：I met a boy…

（2）crying 后面是状语从句 "as if his heart would break"。

（3）"when I spoke to him" 是介于 "said" 和 "that he was hungry because" 之间的插入语。

第三，难点部分的处理："crying as if his heart would break" 应译为"哭得伤心极了"。

译文：在非洲，我遇到一个小孩，他哭得伤心极了，我问他时，他说他饿了，两天没有吃饭了。

一般来说，长句的翻译有顺序法、逆序法、分译法和综合法四种。现将顺序法举例说明如下：

有些英语长句叙述的一连串动作按发生的时间先后安排，或按逻辑关系安排，与汉语的表达方式比较一致，可按原文顺序译出。例如：

Combined with digital television sets, videodiscs can not only present films but also offer surround sound which provides theatre quality-amazing reality by which the viewers may have an illusion that they were at the scene and witnessed everything happening just around them.

分析：按意群的关系，该句可以拆分为五部分：Combined with digital television sets/videodiscs can not only present films but also offer surround sound/which provides theatre quality-amazing reality/by which the viewers may have an illusion/that they were at the scene and witnessed everything happening just around them. 除了必要的增减词，原文各句的逻辑关系，表达次序与汉语基本一致，因此可以按原文译出。

译文：与数字式电视机相结合，图像光盘不仅可以演电影，还提供环境声音，产生电影院效果——令人吃惊的真实感，使观看者产生一种错觉，以为他们在现场目睹他们周围发生的一切。

Section 2 Technology of Improving Transmission Line Capacity and AC/DC Transmission System Stability Level

China's electric load is expected to be increasing steadily and rapidly for a long period. How to improve transmission capacity and AC/DC system stability is becoming one of the major issues. Addressing this issue requires the adoption of new technologies, including compact power transmission technology, series/parallel compensation technology, FACTS technology, such as SVC and TCSC, and high voltage controlled shunt reactor (CSR), etc. In addition to enhancing transmission capacity, these technologies offer ways to improve transient and dynamic stability of power grids, and mitigation of power frequency over voltage. Currently, these technologies have already been applied in China's power grids.

1. Compact transmission line

Compact transmission technology adopts an optimized configuration of conductors to improve the surge impedance loading (SIL) of the line, reduce transmission corridor width, and enhance transmission capacity per unit corridor. When compared with the conventional transmission line, the compact transmission line features better symmetry of phase parameters, smaller positive sequence reactance, higher SIL and smaller corridor width, while maintaining equivalent insulation strength and electric field intensity on conductor surface to ensure a safe condition for liveline work. Currently, the compact transmission technology has been applied in China, and quite a few compact transmission lines have been put into operation.

In September 1994, China's first 220kV compact transmission line, the Anding-Langfang 220kV line was put to use. This 23.95km line starts at the Anding substation in Beijing and ends at the Datun substation in Langfang, Hebei province. The three bundles are arranged in two ways, vertical and inverted triangle. Each phase conductor is composed of four lines, occupying a space of only 1/3rd that of the traditional transmission lines whilst with 1.7 times of SIL.

China's first 330kV compact transmission line was built in October 2004. It is 115km long, at the altitude of 1500~2000m, stretching from the 330kV Chengxian substation at south side to the Tianshui substation at north side. The adoption of compact transmission technology reduced the line corridor considerably from 13~22m to 5.2m, improved electromagnetic environment, enhanced SIL and system stability, and saved the engineering cost per unit transmission capacity. The Fangshan-Changping 500kV line is China's first 500kV compact transmission line which began operation in November 1999. The line extends 83km, with 6-split sub-conductor. Phase conductors are arranged in the inverted-delta mode, with phase spacing of 6.7m and conductor-to-ground distance of 10m, reducing the transmission corridor to 17.9m. The SIL of the transmission line is 1340MW, 34% more than general transmission line.

Besides the above-mentioned lines, other compact transmission lines have also been built, including the Zhengping-Yixin 500kV double circuit compact transmission for the Three Gorges project. Many other projects are currently under construction.

2. Parallel and series compensation

The capacity of a long-distance AC transmission line is generally restricted by its transient stability limit. The introduction of parallel or series compensation capacitor to a transmission line can alleviate such constraints by decreasing the electric distance of the line and improving system stability limit and transmission capacity. These benefits coupled with price advantage, parallel and series compensation equipments have been widely applied in longdistance bulk transmission systems. Currently, a number of series compensation equipments have been installed in China, including the Sanbao SC project in Jiangsu province for the Yangcheng Power Plant, the Weixian SC Project at the Datong-Fangshan double circuit transmission line in the North China Power Grid, the Wanquan SC project at the Fengzheng-Wanquan-Shunyi double circuit transmission line, the Pingguo SC project at the Tianshengqiao-Guangzhou double circuit transmission line in the South China Power Grid and the Hechi SC at the Huishui-Hechi double circuit transmission line. More series compensation equipments are expected to be installed in the transmission lines in the future.

3. SVC, TCSC, and other FACTS equipments

For AC or AC and DC parallel long distance bulk transmission corridors, the power grid's dynamic reactive power supporting capability under major fault disturbance not only restricts the transmission capacity of the interconnected network but also impacts the voltage stability level and safety margin of the interconnected systems. Using the dynamic reactive compensation, it is possible to regulate the reactive power rapidly in response to the system needs and maintain the bus voltage around rated value. SVC has been widely utilized as a dynamic reactive compensation technology.

The application of SVC in the power grid can quickly change the reactive power generation, providing dynamic reactive power supply for power systems, and regulating system voltage. In 2006, three SVC equipments with a total capacity of 420Mvar were put into operation at the east corridor of Chuanyu Power Grid, with one 120Mvar SVC installed at the Chenjiaqiao 500kV substation, one 120Mvar SVC installed at the Honggou 500kV substation, and one 180Mvar SVC at the Wanxian 500kV substation. The transmission capability of Chuanyu network interface was improved by 300MW and damping ratio of oscillation increased as well. As a result, the system's transient and dynamic stability was considerably improved. Additionally, the corridor's fourth set of SVC, 500kV Yongchuan substation SVC, rated at 120Mvar, is currently under construction. At present, China has completed the localization of a large capacity SVC for 500kV transmission systems that costs around RMB 300Yuan/kvar.

The thyristor controlled series compensation technology (TCSC) is a new FACTS technology developed in recent years. In addition to the technical advantages over the traditional series compensation, TCSC features a rapid controllability to further improve system stability and transmission capacity, damp low frequency oscillation, eliminate SSR, and optimize system operation mode. This makes TCSC a highly effective means of improving the capacity of transmission corridors.

4. High voltage CSR

CSR resolves the conflict between over voltage limitation and reactive power compensation. The installation of CSR at a long distance extra high-voltage line can not only flexibly adjust the system reactive power according to different operating modes and ensure the voltage safety of transmission line at various times; but also reduce network losses, improve network dynamic stability in certain degree, and increase transmission capacity.

CSR can be categorized into two types based upon the magnetic control principle and high impedance transformer principle, respectively. Currently, two sets of CSR have been applied in China.

The Jingzhou CSR project employed a 500kV magnetic controlled CSR for the first time in China. The reactor located at the Jianglin side of the Three Gorges right bank to the Jianglin line (the 130km Three Gorges right bank to the Jianglin single circuit line is part of the Three Gorges transmission and substation project) has a designed capacity of 3×40Mvar. Two sets of 100Mvar reactors, one fixed and one controlled will be installed at the right bank of the Three Gorges and the Jianglin side, respectively. The magnetic controlled reactor enables the effective regulation of systemvoltage by rapidly and continuously adjusting reactance and capacity.

Furthermore, in case of emergency this CSR can offer forced compensation to mitigate the power frequency over voltage. CSR in conjunction with neutral reactor can restrain secondary arc current and decrease recovery voltage to ensure the success rate of singlephase re-closure in order to improve the reliability of transmission system. This project is currently under construction.

5. High speed protection and special control systems

Based on the design principal of "Three Defense Lines", advanced automation systems, relay protection devices, and special protection system have ensured the security of China's power grid under the condition of relatively weak grid structure.

At present, these high-speed protections have showed good performance in reliability and fast response. From the perspective of reliability, in 2004, the ratio of correct protection in 220kV or above lines and transformers is 99.21%; the ratio of fast response is 99.33%, and 96.04% protection devices are micro-computer based protections. From the perspective of fast response, the fault clearing time of relay protection used in 220kV or above lines is 30~50ms. In addition, new technologies of ultra high-speed protection devices are being adopted in order to meet the need of EHVAC and UHVAC power grids.

In addition to high-speed protection systems, inter-regional security and stability control systems, lighting strike spot location systems, wide area measurement systems, and fault location systems for transmission lines are widely used in China's power grids. These systems have ensured the security and stability of China's power system, and lower the risk of large-area blackouts. In the future, China will conduct research on security protection systems with abilities of self-adaptability, inter-regional coordination and forecast, real-time dynamic security assessment technology, real-time data platform combining functions of SCADS/EMS and WAMS, new EMS with abilities of real time monitoring, dynamic security assessment and pre-warning, auxiliary decision making, automatic voltage control systems, etc.

原文翻译

第二节 提高输电线路容量的技术与交/直流输电系统的稳定性水平

中国的电力负荷预计会长时间平稳、较快地增加，如何提高传输容量和交/直流系统的稳定性成为关键问题。解决这一问题需要采用新的技术，包括紧凑型输电技术、串/并联补偿技术、柔性交流输电技术，如静止无功补偿（SVC）和可控串联补偿装（TCSC），以及高电压控制并联电抗器（CSR）等。除了提高传输容量，这些技术还可以用来提高电网的暂态和动态稳定，减小工频过电压。目前，这些技术已在我国电网中应用。

1. 紧凑型输电线路

紧凑型输电技术采用优化配置来提高线路自然功率（SIL），减少输电走廊宽度，提高单位走廊上的传输容量。与常规线路相比，紧凑型输电线路的相参数对称性较好，正序电抗小，有更高的特性阻抗加载和较小的走廊宽度，并且能维持相同的绝缘强度和导体表面的电场强度，这样可以确保带电作业的安全。目前，紧凑型输电技术已应用在中国，并且不少紧凑型输电线路已投入运行。

1994年9月，中国第一条220kV紧凑型输电线路——安定—廊坊220kV线路投入使用。这条长度为23.95km的线路从北京安定变电站开始，在河北省的廊坊大屯变电站结束。三相导线的排列方式有2种，即垂直和倒三角形。每相导线是由四条线组成，占用的空间只有传统传输线的1/3，为线路自然功率的1.7倍。中国首条330kV紧凑型输电线路建于2004年10月，其长达115km，海拔达1500~2000m，从南侧的330kV成县变电站一直延伸到北侧的天水变电站。采用紧凑型输电技术可以将 13～22m 的线路走廊骤减为 5.2m，改善电磁环境，增强线路自然功率和系统的稳定性，并且可以降低单位输电容量的工程费用。500kV 房山—昌平输电线路是我国第一条紧凑型 500kV 输电线路，于 1999 年 11 月开始运行。该线路长达 83km，采用 6 分裂导线。相导线布置为倒三角模式，导线距地面的距离分别为 6.7m 和 10m，输电走廊减小到了 17.9m。该输电线路的线路自然功率是 1340MW，比一般输电线路多 34%。

除了上述线路，其他紧凑型输电线路也已建成，包括三峡工程中的 500kV 郑平—沂新双回紧凑型输电线路。目前有许多其他项目也正在建设。

2. 并联和串联补偿

长距离交流输电线路的容量通常受其暂态稳定极限的限制。输电线路引进并联或串联补偿电容器可以通过降低导线的电气距离来减小这些限制，并提高系统的稳定性及输送容量。由于这些好处以及价格优势，并联和串联补偿装置已被广泛应用于远距离输电系统。目前，中国已安装了有大量成系列的补偿设备，包括江苏省阳城电厂的三宝串联电容补偿（SC）项目，华北电网大同—房山双回路输电线路、丰镇—万全—顺义双回路输电线路的串联电容补偿项目、天生桥—广州双回路输电线路和惠水—河池双回路输电线路串联电容补偿项目。预计未来会安装更多的串联补偿输电线路设备。

3. 静止无功补偿装置、可控串联补偿装置和其他灵活交流输电设备

对于交流或交流和直流并行长距离输电走廊，电网在严重故障干扰下的动态无功功率支持容量不仅限制了电网的传输容量，而且还影响了互连系统的电压稳定性和安全裕度。使用动态无功补偿装置就可以迅速响应系统需要的无功功率的调节和维持母线电压在额定值附近。静止无功补偿装置已被广泛用作动态无功补偿技术。

在电力系统中的静止无功补偿装置可以快速更改无功功率，为电力系统提供动态无功功率和调节系统电压。2006 年，三个总容量为 420Mvar 的静止无功补偿装置在川渝电网的东部投入运行，一个 120Mvar 静止无功补偿装置安装在 500kV 陈家变电站，一个 120Mvar 静止无功补偿装置安装在 500kV 洪沟变电站，一个 180Mvar 静止无功补偿装置安装在 500kV 万县变电站。川渝电网接口的传输容量提高了 300MW，振荡的阻尼比也有所提高。因此，系统的瞬态和动态稳定性大大改善。此外，东部电网的第四套静止无功补偿装置，额定功率为 120MW 的 500kV 永川变电站静止无功补偿装置目前正在建设中。目前，中国已经为 500kV 输电系统完成了一个大容量静止无功补偿装置的安装，费用约为每千乏 300 元。

可控串联补偿技术（TCSC）是近年来发展起来的一种新型灵活交流输电技术。除了传统的串联补偿技术优势，可控串联补偿技术具有快速可控性，以进一步提高系统的稳定性和传输能力，抑制低频振荡，消除固态继电器和优化系统运行模式。这是可控串联补偿技术提高传输能力的一个非常有效的手段。

4. 高压可控并联电抗器

可控并联电抗器（CSR）解决了电压限制和无功功率补偿之间的冲突。安装在远距离超高压线路的可控并联电抗器不仅可以根据不同的运行模式和不同的时间，灵活地调整系统无功功率，确保输电线路的电压安全，同时还可以降低网络损耗，在一定程度上提高电网动态稳定，增加传输容量。

可控并联电抗器根据磁控制原理和高阻抗变换器原理可分为两种类型。目前，这两种类型的设备都已经在中国应用。

荆州可控并联电抗器项目是中国第一个采用 500kV 磁控制可控并联电抗器的项目。位于三峡右岸江林线靠近江林（长达 130km 的三峡右岸向江林单回线路是三峡输变电工程的一部分）的电抗器，其设计功率 3×40Mvar。两套 100Mvar 反应堆，另一个固定控制将分别安装在三峡、江林。磁性的可控电抗器有通过迅速和不断调整电抗使系统电压得到有效监管的能力。

此外，在紧急情况下，这种可控并联电抗器可以提供强制补偿来减轻工频过电压。结合中性点小电抗的可控并联电抗器可以抑制潜供电流和恢复低电压，这样可以确保单相重合闸的成功率，以提高输电系统可靠性。这个项目正在建设中。

5. 高速保护和特殊控制系统

基于设计当中主要的"三个防线"，先进的自动化系统、继电保护装置和特殊的保护系统保证了中国相对脆弱的电网结构下的电网安全。

目前，这些高速保护装置在可靠性和快速响应方面均表现出良好的性能。从可靠性的角度来看，2004 年中 220kV 或以上线路和变压器的正确保护率是 99.21%，快速响应的正确率是 99.33%，其中 96.04%的保护装置是微机保护。从快速响应的角度来看，继电保护装置在 220kV 及以上线路故障切除时间为 30~50ms。此外，可以采用超高速保护装置新技术以满足超高压交流和特高压电网的需要。

除了高速保护系统，区域间安全和稳定控制系统、雷击点定位系统、广域测量系统和输电线路故障定位系统等也在中国电网中被广泛使用。这些系统保证了中国电力系统的安全性和稳定性，并降低了大面积停电风险。在未来，中国将开展具有自适应能力的安全保护系统研究，这个系统具有区域间的协调和预测、实时动态安全评估技术、结合监控系统/能量管理系统和广域测量的实时数据平台的功能，也会研究新的能量管理系统，该系统具有实时监控、动态安全评估和预警、制订备用方案和自动电压控制系统等功能。

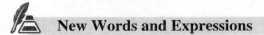

New Words and Expressions

series/parallel compensation	串/并联补偿
thyristor	*n.* ［电子］半导体闸流管
electric load	电负载，电气荷载
surge impedance loading	自然功率
phase parameter	相参数
positive sequence	正序
insulation strength	绝缘强度
series compensation	［电子］串联补偿
damping ratio	阻尼系数；阻尼比率；衰减率
singlephase	单相

Notes

1. Compact transmission technology adopts an optimized configuration of conductors to

improve the surge impedance loading (SIL) of the line, reduce transmission corridor width, and enhance transmission capacity per unit corridor.

紧凑型输电技术采用优化配置来提高线路自然功率（SIL），减少输电走廊宽度，提高单位走廊上的传输容量。

2. Using the dynamic reactive compensation, it is possible to regulate the reactive power rapidly in response to the system needs and maintain the bus voltage around rated value.

使用动态无功补偿装置就可以迅速响应系统需要的无功功率的调节和维持母线电压在额定值附近。

3. Based on the design principal of "Three Defense Lines", advanced automation systems, relay protection devices, and special protection system have ensured the security of China's power grid under the condition of relatively weak grid structure.

基于设计当中主要的"三个防线"，先进的自动化系统、继电保护装置和特殊的保护系统保证了中国相对脆弱的电网结构下的电网安全。

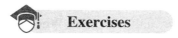
Exercises

Ⅰ. Choose the best answer into the blank.

1. The _____ 500kV line is China's first 500kV compact transmission line which began operation in November 1999.
  A. Chengxian-Tianshui   B. Anding-Langfang
  C. Fangshan-Changping   D. Zhengping-Yixin

2. The introduction of_____ to a transmission line can alleviate.
  A. Compact transmission technology   B. parallel or series compensation capacitor
  C. High voltage CSR   D. special protection system

3. The application of SVC in the power grid can quickly change _____.
  A. the reactive power   B. the active power
  C. apparent power   D. total power

4. _____ resolves the conflict between over voltage limitation and reactive power compensation.
  A. SVC   B. CSR   C. TCSC   D. SIL

5. High-speed protections have showed good performance in _____ and fast response.
  A. reliability   B. economy   C. security   D. sensitivity

Ⅱ. Answer the following questions according to the text.

1. What is the role of conductor of compact transmission technology?

2. What problems can be solved by the CRS?

Ⅲ. Translate the following into chinese.

In addition to high-speed protection systems, inter-regional security and stability control systems, lighting strike spot location systems, wide area measurement systems, and fault location systems for transmission lines are widely used in China's power grids. These systems have ensured the security and stability of China's power system, and lower the risk of large-area blackouts. In the future, China will conduct research on security protection systems with abilities of self-adaptability, inter-regional coordination and forecast, real-time dynamic security assessment technology, real-time data platform combining functions of SCADS/EMS and WAMS, new EMS with abilities of real time monitoring, dynamic security assessment and pre-warning, auxiliary decision making, automatic voltage control systems, etc.

Keys

Ⅰ. Choose the best answer into the blank.
1. C　　2. B　　3. A　　4. B　　5. A

Ⅱ. Answer the following questions according to the text.

1. Compact transmission technology adopts an optimized configuration of conductors to improve the surge impedance loading (SIL) of the line, reduce transmission corridor width, and enhance transmission capacity per unit corridor. When compared with the conventional transmission line, the compact transmission line features better symmetry of phase parameters, smaller positive sequence reactance, higher SIL and smaller corridor width, while maintaining equivalent insulation strength and electric field intensity on conductor surface to ensure a safe condition for liveline work.

2. CSR resolves the conflict between over voltage limitation and reactive power compensation.

Ⅲ. Translate the following into chinese.

除了高速保护系统，区域间安全和稳定控制系统、雷击点定位系统、广域测量系统和输电线路故障定位系统等也在中国电网中被广泛使用。这些系统保证了中国电力系统的安全性和稳定性，并降低了大面积停电风险。在未来，中国将开展具有自适应能力的安全保护系统研究，这个系统具有区域间的协调和预测、实时动态安全评估技术、结合监控系统/能量管理系统和广域测量的实时数据平台的功能，也会研究新的能量管理系统，该系统具有实时监控、动态安全评估和预警、制订备用方案和自动电压控制系统等功能。

翻译技巧之长句（三）

长句的翻译方法中关于逆序法的举例说明如下：

"逆序法"又称"倒置法"，主要指句子的前后倒置问题。有些英语长句的表达次序与汉语习惯不同，甚至语序完全相反，这就必须从原文的后面译起，逆着原文的顺序翻译。逆序法在长句的翻译中，我们可根据不同的情况按意群进行全部逆序或部分逆序。例如：

There are many wonderful stories to tell about the places I visited and the people I met.

分析：该句可以拆分为三部分：There are many wonderful stories to tell about/the places I visited/the people I met.这句英语长句的叙述层次与汉语逻辑相反，所以采用逆译法。

译文：我访问了一些地方，遇到不少人，要谈起来，奇妙的事可多着呢。

There was little hope of continuing my inquiries after dark to any useful purpose in a neighborhood that was strange to me.

分析：该句可分为三部分:There was little hope/continuing my inquiries after dark to any useful purpose/in a neighborhood that was strange to me.第一、二层次是表示结果，第三层次表示原因，按照中文的表达习惯，常把原因放在结果前面。这句英语长句的叙述层次与汉语逻辑相反，因此要打破原句的结构，按照汉语造句的规律重新加以安排。

译文：这一带我不熟悉，天黑以后继续进行调查，取得结果的希望不大。

Section 3 Power Grids and Grid Interconnections in China

1. Provincial and big regional power grids

After years of development, six inter-provincial regional power grids, including the Northeast China Power Grid, the North China Power Grid, the Central China Power Grid, the East China Power Grid, and the Northwest China Power Grid as well as three independent provincial grids, Xinjiang, Tibet, Hainan power grids have been formed in the main land of China.

2. Interconnection of regional grids by AC or DC lines

The distribution of fossil and water resources in China is not even. Most of these natural resources locate in the west China. For example, 274GW water energy resource or 72% of China's total utilizable water energy resource and 388.2 billion ton coal or 39% of China's coal reserve are in western China. However, most major load centers of China are located in the east. Therefore, in order to make resources to be used more efficiently, China has made a strategy of transferring power from the west to the east, and creating a nationwide power grid interconnection.

The interconnection of China's regional grids started in the late 1980s. By the end of 2005, China had mostly completed the interconnection of large regional power grids and built an ultra large-scale interconnected power grid across all provinces in the main land China except Xinjing, Tibet, and Hainan. The installed capacity of the interconnected power grid reached 400GW. More detailed information about the interconnections of Chinese large regional grids in 2005 is in Table8.1.

Table8.1　　　The interconnections of Chinese large regional grids in 2005

Interconnection	AC lines	DC lines
Northeast China power grid-North China Power Grid	Double circuit lines from Gaoling to Jiangjiayin	—
North Power Grid-Central China Power Grid	Single circuit line from Xin'an to Huojia	—
Central China Power Grid-East China Power Grid	—	Gezhouba-Nanqiao; Longquan-Zhengpin
Central China Power Grid-South China Power Grid	—	Jianglin-E'Cheng
Central China Power Grid-Northwest China Power Grid	—	Zidong-Lingbao-Luofu back to back

In the mean time, China had built three power transmission corridors, i.e., the north, the center, and the south corridors, to transmit power from the west to the east. The north corridor is from the Northwest (including Xinjiang) China Power Grid to the North (including Shandong province) China Power Grid. The center corridor is from the Central China (including Chuanyu) Power Grid to the East China (including Fujian) Power Grid. The south corridor is inside the South China Power Grid. In the summer of 2005, the general power flow of the four sub-corridors reached 10700MW. The center corridor includes a 1200MW DC line from Gezhou Dam to Shanghai and a 3000MW Longzhen DC line from the Three Gorges to Changzhou, Jiangsu province. The south corridor includes nine lines (six AC lines and three DC lines), i.e., four 500kV AC lines from Tianshenqiao to Guangdong province, two 500kV AC lines from Guizhou province to Guangdong province, a ±500kV DC line from Tianshengqiao to Guangdong, a ±500kV DC line from Guizhou to Guangdong, a ±500kV DC line from the Three Gorges to Guangdong. In 2005, the capacity from several southwestern provinces to Guangdong province was 10880MW. In the future, China will continue to improve the interconnection of the regional grids. Initiatives include the construction of two 1000kV power transmission corridors from northern China to southern China and from western China to eastern China, and a back-to-back DC project interconnecting the Northeast China Power Grid and the North China Power Grid and so on.

 原文翻译

第三节 我国电网与电网互联情况

1. 省网和大区域电网

经过多年的建设发展，全国共有 6 个互联的区域电网和 3 个独立的省网。区域电网分别是东北电网、华北电网、华中电网、华东电网、西北电网，省网分别为新疆、西藏和海南电网。

2. 区域电网的交直流互联

中国的化石能源和水力资源分布是非常不均衡的，主要集中在西部地区。例如，在西部地区，水力资源蕴藏量达 274GW，其中可利用的资源约有 72%；而煤炭蕴藏量达 3882 亿吨，其中可利用的资源约占总资源的 39%。然而，我国的负荷中心主要集中在东部地区，因此，为了更有效地利用能源资源，我国已经制订了一个由将能源从西部输往东部的发展战略，并逐步形成全国电网互联。

20 世纪 80 年代末期，中国开始发展区域电网互联。截至 2005 年年底，我国基本上完成了全国各大区域电网互联，并在除新疆、西藏和海南之外的地区建立了一个横跨各省的特大型互联电网，其总装机容量达到 400GW，有关 2005 年我国大区域电网互联信息的详细情况，见表 8.1。

表 8.1　　2005 年我国大区域电网互联情况

互联电网	交流输电线路	直流输电线路
西北电网—华北电网	高陵—姜家垴双回输电线路	—
华北电网—华中电网	新安—霍家单回输电线路	—
华中电网—华东电网	—	葛洲坝—南桥；龙泉—征品
华中电网—南方电网	—	江陵—鹅城
华中电网—西北电网	—	梓东—岭鄢—罗福背靠背工程

与此同时，为了实现西电东送，我国已经建成了北、中和南等三个电能输送通道走廊。北通道是将电能从西北电网（含新疆）输送到华北电网（含山东省），中通道是从华中电网（包括四川省和重庆市）到华东电网（包括福建省），而南通道则位于中国南方电网内部。2005年夏天，4个分支走廊输送的电力总功率达到 10 700MW。中通道包括 2 项直流输电工程，分别是葛洲坝—上海的额定容量为 1200MW 直流输电线路和三峡—江苏常州的额定容量为 3000MW 龙镇直流输电线路。南通道包括 9 条高压输电工程（6 条交流输电线路和 3 条直流输电线路），分别为 4 条从天生桥—广东 500kV 交流输电线路，2 条贵州—广东 500kV 交流输电线路，1 条从天生桥—广东的±500kV 直流输电线路，1 条从贵州—广东的±500kV 直流输电线路，1 条从三峡—广东的±500kV 直流输电线路。2005 年，从我国西南地区输送到广东的电能容量达到了 10 880MW。未来几年，我国将会陆续完善各区域电网之间的互联状况，初步建设规划是：2 条 1000kV 电压等级的从北部到华南和西部到东部的输电通道走廊，以及 1 项用于实现东北电网和华北电网同步互联的背靠背直流输电工程等。

New Words and Expressions

capacity　　　　　　　　　　　　　　　　　　n. 能力、容量

Notes

1. After years of development, six inter-provincial regional power grids, including the Northeast China Power Grid, the North China Power Grid, the Central China Power Grid, the East China Power Grid, and the Northwest China Power Grid as well as three independent provincial grids, Xinjiang, Tibet, Hainan power grids have been formed in the main land of China.

经过多年的建设发展，全国共有 6 个互联的区域电网和 3 个独立的省网。区域电网分别是东北电网、华北电网、华中电网、华东电网、西北电网和南方电网，省网分别为新疆、西藏和海南电网。

2. Therefored, in order to make resources to be used more efficiently, China has made a strategy of transferring power from the west to the east, and creating a nationwide power grid interconnection.

因此，为了更有效地利用能源资源，我国已经制订了一个由将能源从西部输往东部的发

展战略,并逐步形成全国互联电网。

3. Initiatives include the construction of two 1000kV power transmission corridors from northern China to southern China and from western China to eastern China, and a back-to-back DC project interconnecting the Northeast China Power Grid and the North China Power Grid and so on.

初步建设规划是:2条1000kV电压等级的从北部到华南和西部到东部的输电通道走廊,以及1项用于实现东北电网和华北电网同步互联的背靠背直流输电工程等。

Exercises

Ⅰ. Choose the best answer into the blank.

1. The distribution of fossil and water resources in China is ____.
   A. extensive   B. even   C. scanty   D. nonuniform
2. Natural resources locate in the ____ west.
   A. east   B. south   C. west   D. north
3. The installed capacity of the interconnected power grid reached ____ GW.
   A. 274   B. 400   C. 3000   D. 10 880
4. China had built ____ power transmission corridors.
   A. one   B. two   C. three   D. four

Ⅱ. Answer the following questions according to the text.

1. What does the provincial and big regional power grids consist of?
2. Why does china make a strategy of transferring power from the west to the east?
3. Which the transmission corridor has been established in China?

Ⅲ. Translate the following into Chinese.

The distribution of fossil and water resources in China is not even. Most of these natural resources locate in the west China. For example, 274GW water energy resource or 72% of China's total utilizable water energy resource and 388.2 billion ton coal or 39% of China's coal reserve are in western China. However, most major load centers of China are located in the east. Therefore, in order to make resources to be used more efficiently, China has made a strategy of transferring power from the west to the east, and creating a nationwide power grid interconnection.

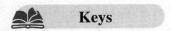

Keys

Ⅰ. Choose the best answer into the blank.

1. D   2. C   3. B   4. C

Ⅱ. Answer the following questions according to the text.

1. After years of development, six inter-provincial regional power grids, including the Northeast China Power Grid, the North China Power Grid, the Central China Power Grid, the

East China Power Grid, and the Northwest China Power Grid as well as three independent provincial grids, Xinjiang, Tibet, Hainan power grids have been formed in the main land of China.

2. In order to make resources to be used more efficiently, China has made a strategy of transferring power from the west to the east, and creating a nationwide power grid interconnection.

3. China had built three power transmission corridors, i.e., the north, the center, and the south corridors, to transmit power from the west to the east.

III. Translate the following into Chinese.

我国的化石能源和水力资源的分布是非常不均衡的，主要集中在西部地区。例如，在西部地区，水力资源蕴藏量达 274GW，其中可利用的资源约有 72%；而煤炭蕴藏量达 3882 亿吨，其中可利用的资源约占总资源的 39%。然而，我国主要负荷中心主要集中在东部地区，因此，为了更有效地利用能源资源，我国已经制订了一个由将能源从西部输往东部的发展战略，并逐步形成全国互联电网。

翻译技巧之长句（四）

长句的翻译方法中关于分译法的举例说明如下：

"分译法"又叫"拆译法"，有些英语长句的主句与从句或主句与修饰语间的关系并不十分密切，可把长句中的从句或短语化为句子分开来叙述，这符合汉语多用短句的习惯。为使意思连贯，有时还可适当增加词语。考生必须注意的是，分译的目的是化长为短、化整为零，消除译文的阻塞；分译后的译文必须连贯，有整体感。例如：

He became deaf at five after an attack of typhoid fever.

分析：这个英语句子，有两个介词短语，代表两层意思，表示什么时候，生了什么病。翻译时打破原句的结构，按汉语造句的规律重新安排。

参考译文：他五岁的时候，生了一场伤寒病，变成了聋子。

I was on my way home from tramping about the streets, my drawings under my arm, when I found myself in front of the Mathews Gallery.

分析：这句英语长句的叙述层次与汉语逻辑相反，因此要打破原句的结构，按照汉语造句的规律重新加以安排。

参考译文：我夹着画稿，在街上兜了一番，回家的路上无意中发现自己逛到了马太画廊的门口。

Chapter 9 The Development of China Smart Grid

Section 1 How Green Is The Smart Grid?

A simulation of the US power system suggests that both conservative and more technologically aggressive implementations of a smart grid would produce a significant reduction in power sector carbon emissions at the national level. A conservative approach could reduce annual CO_2 emissions by 5 percent by 2030, while the more aggressive approach could lead to a reduction of nearly 16 percent by 2030.

In any long-term vision for the power industry one is likely to come across two words: smart, grid. With an aging electric infrastructure and a renewed national interest in energy and climate issues, policymakers and industry leaders are recognizing the significant potential that an infrastructure upgrade accompanied by new, cutting-edge demand-side technologies can play in the nation's portfolio of energy resources. The potential benefits of the smart grid have been cited widely. Most notably, they are highlighted in President Barack Obama's new energy plan, which calls for aggressive investment in smart grid technologies.

Often included in the enumeration of smart grid benefits are improvements to the environment, particularly in the form of lower greenhouse gas emissions. The smart grid is front and center in the debate on climate change and former Vice President Al Gore has called for a $400 billion investment to modernize the grid in his campaign against global warming. However, very little research has been published that quantifies these potential environmental benefits.

This article seeks to redress this gap. Given the amorphous definition of the smart grid at this early stage in its development, two scenarios are examined. The first scenario considers the impacts of an upgrade to technologies that are commercially available today, including advanced metering infrastructure, dynamic pricing, automating technologies, and information displays. The second scenario takes an expanded view of the smart grid to include the possible impacts of future technologies that could become available in the long-term. This includes the impacts of a smart distribution system and an increase in distributed energy resources.

In estimating the potential environmental impacts of the smart grid, it is critical to account for regional system dynamics. Each region of the United States has a very different mix of electricity generation resources. For example, the Pacific Northwest relies largely on hydropower, while the Midwest gets much of its electricity from coal plants. These resources all have different CO_2 emissions rates and are utilized at different times of the day. As a result, the timing and nature of the smart grid's impact on electricity consumption will have a significant influence on the ultimate reduction in carbon emissions. In this study, a modeling system called the Regional Capacity Planning model (RECAP) is used to capture these complex relationships.

Ultimately, in both the Conservative and Expanded Scenarios, the smart grid is found to produce a significant reduction in power sector carbon emissions at the national level. A conservative approach to the smart grid could reduce annual power sector CO_2 emissions by 5 percent by 2030, while the more aggressive pursuit of a broader range of technologies could lead to reduction of nearly 16 percent in annual emissions by 2030.

 原文翻译

第一节　智能电网有多环保？

美国的电力系统仿真实验表明，在国家层面上实施保守或更积极的智能电网技术将使碳排放量大幅度减少。到 2030 年，使用保守的技术可以每年减少 5% 的 CO_2 排放量，而使用更积极的技术可以使 CO_2 每年排放量减少约 16%。

从电力行业的长远角度来看，我们很可能遇到两个词：智能、电网。伴随着逐渐老化的电力基础设施与国家重心向能源和气候问题的转变，政府决策者和业界领导者意识到，伴随着新的尖端需求侧技术的发展，基础设施也随之更新，这在国家能源资源的投资组合中可能发挥重大潜能。智能电网的潜在优势被广泛引用。值得强调的是，这些优势在呼吁向智能电网积极投资的巴拉克·欧巴马总统的新能源计划中被着重强调。

智能电网的优势之一是改善环境，特别是在降低温室气体排放方面。智能电网在关于气候变化的讨论中往往是前沿和中心问题，前副总统阿尔·戈尔在抵制全球变暖的活动中呼吁拿出 4000 亿美元来投资建设更具现代化的电网。然而，几乎没有关于潜在环境效益如何量化的研究成果发表。

这篇文章旨在弥补这方面空白。鉴于智能电网在其发展早期的定义不明确，可从两种情况对其进行研究。第一种方案考虑升级现今已有技术，包括先进的计量基础设施、动态定价、自动化技术和显示信息的影响。第二种方案是以发展的眼光看待智能电网，包括未来可能实现的并且在今后会长期使用的。这包括分布式资源的增加和智能配电系统的影响。

为估计智能电网对环境的潜在影响，关键是要考虑区域系统动态。美国每一个区域都有不同的资源。例如，太平洋西北部主要依赖于水力发电，而美国中西部的电能主要依靠火力发电厂。所有这些资源都有不同程度的二氧化碳释放率，而且每天的排放次数不尽相同。最后得到的结论是，影响智能电网电力消耗的时间特性及性质将对二氧化碳排放量有非常重大的影响。在该项研究中，地域性容量规划模型（RECAP）模型系统被用来描述不同变量间的复杂关系。

最后，在保守和扩展的方案中，智能电网可以大规模减少在国家层面上的电力单元产生的二氧化碳排放量。到 2030 年时，智能电网采取保守方案时估计能够每年减少电力单元产生的二氧化碳排放量的 5%，而采用更广泛先进技术的话能够每年减少 16%。

New Words and Expressions

redress	v. 补偿，弥补
electricity generation resources	发电资源
smart grid	智能电网
electricity consumption	用电，电力消耗

Notes

1. A simulation of the US power system suggests that both conservative and more technologically aggressive implementations of a smart grid would produce a significant reduction in power sector carbon emissions at the national level.

美国电力系统的一项仿真表明，对于智能电网中相对保守的和较为积极的技术的实施将在国家范围内使碳排放量大幅度减少。

2. With an aging electric infrastructure and a renewed national interest in energy and climate issues, policymakers and industry leaders are recognizing the significant potential that an infrastructure upgrade accompanied by new, cutting-edge demand-side technologies can play in the nation's portfolio of energy resources.

伴随着逐渐老化的电力基础设施与国家重心向能源和气候问题的转变，政府决策者和业界领导者意识到伴随着新的前沿需求侧技术的基础设施更新在国家能源资源的投资组合中具有重大潜能。

3. Ultimately, in both the Conservative and Expanded Scenarios, the smart grid is found to produce a significant reduction in power sector carbon emissions at the national level.

最后，在保守和扩展的方案中，智能电网被发现可以大规模地减少在国家层面上的电力单元产生的二氧化碳排放量。

Exercises

Ⅰ. Choose the best answer into the blank.

1. A simulation of the US power system suggests that both ＿＿ and more technologically ＿＿ implementations of a smart grid would produce a significant reduction in power sector carbon emissions at the national level.

　　A. conservative;negative　　　　B. onservative;aggressive
　　C. negative;aggressive　　　　　D. aggressive;conservative

2. Often included in the enumeration of smart grid benefits are improvements to the ＿＿, particularly in the form of lower greenhouse gas emissions.

　　A. environment　　B. smart grid　　C. government　　D. nation

3. In any long-term vision for the ＿＿one is likely to come across two words: "smart

grid."

    A. energy industry   B. electric industry

    C. power industry   D. environmental organization

4. The smart grid is found to produce a significant reduction in power sector carbon emissions at the _____ level.

    A. national   B. regional   C. world   D. community

5. The first scenario considers the impacts of an upgrade to technologies that are commercially available today, including advanced metering infrastructure, dynamic pricing, _____ technologies, and information displays.

    A. motivating   B. intelligent   C. artificial   D. automating

II. Answer the following questions according to the text.

1. what kind of implementations of smart grid would lead to a significant reduction in power sector carbon emissions?

2. What makes the policymakers and industry leaders recognize the potential of infrastructure upgrade?

3. What kind of thing does great effect on CO_2 emissions?

III. Translate the following into Chinese.

In any long-term vision for the power industry one is likely to come across two words: smart grid. With an aging electric infrastructure and a renewed national interest in energy and climate issues, policymakers and industry leaders are recognizing the significant potential that an infrastructure upgrade accompanied by new, cutting-edge demand-side technologies can play in the nation's portfolio of energy resources. The potential benefits of the smart grid have been cited widely. Most notably, they are highlighted in President Barack Obama's new energy plan, which calls for aggressive investment in smart grid technologies.

Keys

I. Choose the best answer into the blank.

1.B   2.A   3.C   4.A   5.D

II. Answer the following questions according to the text.

1. A simulation of the US power system suggests that both conservative and more technologically aggressive implementations of a smart grid would produce a significant reduction in power sector carbon emissions at the national level.

2. With an aging electric infrastructure and a renewed national interest in energy and climate issues, policymakers and industry leaders are recognizing the significant potential that an infrastructure upgrade accompanied by new, cutting-edge demand-side technologies can play in the nation's portfolio of energy resources.

3. As a result, the timing and nature of the smart grid's impact on electricity consumption will have a significant influence on the ultimate reduction in carbon emissions.

III. Translate the following into Chinese.

从电力工业的长远角度来看，我们很可能遇到两个词：智能电网。伴随着逐渐老化的电力基础设施与国家重心向能源和气候问题的转变，政府决策者和业界领导者意识到伴随着新的前沿需求侧技术的基础设施更新在国家能源资源的投资组合中具有重大潜能。智能电网的潜在好处被广泛引用。最值得注意的是，这些优势在呼吁向智能电网积极投资的奥巴马总统的新能源计划中非常突出。

翻译技巧之长句（五）

长句的翻译方法中关于综合法的举例说明如下：

有些英语长句顺译或逆译都不合适，分译也有困难，单纯的使用一种译法不能译出地道的汉语时，就应该仔细推敲，或按时间先后，或按逻辑先后，或按逻辑顺序，有顺有逆地进行综合处理。例如：

When Hetty read Arthur's letter she gave way to despair. Then, by one of those conclusive, motionless actions by which the wretched leap from temporary sorrow to life-long misery, she determined to marry Adam.

分析：该句可以分解为五个部分：When Hetty read Arthur's letter/she gave way to despair/by one of those conclusive, motionless actions/by which the wretched leap from temporary sorrow to life long misery/she determined to marry Adam. 首先，第四部分是修饰第三部分的，对第四部分采用"分译法"，然后再适当地根据汉语的表达习惯调整语序，增减词语，对句子进行综合处理。

参考译文：海蒂读过阿瑟的信后陷入绝望之中。后来，她采取了一个不幸者常常采用的那种快刀斩乱麻的行动:嫁给了亚当，从而陷入了终身的痛苦，为的不过是跳出那一时的悲伤。

语序调整是保证句子通顺的关键所在。由于长句的翻译大多需要将原句拆分成多个较短的句子，并且要通过多个短句表达出原来长句的修饰关系和意思，汉语句子的逻辑关系在很大程度上依靠各语法单位在句中的顺序得以表现。所以语序的调整是长句翻译中极其重要的一个环节，直接关系到意思是否完整，句子是否通顺。语序调整是各类词语翻译方法和各种特殊句型翻译的综合运用。

Section 2 Defining The Smart Grid

"Smart grid" means many things to many people. However, there are some consistent themes across the wide range of existing definitions. At a basic level, the smart grid will serve as the information technology backbone that enables widespread penetration of new technologies that today's electrical grid cannot support. These new technologies include cutting-edge advancements in metering, transmission, distribution, and electricity storage technology, as well as providing new information and flexibility to both consumers and providers of electricity. Ultimately, access to this information will improve the products and services that are offered to consumers, leading to more efficient consumption and provision of electricity.

At the core of all smart grid definitions is advanced metering infrastructure (AMI). This refers to smart meters and an accompanying communications network that allows for two-way communication between the provider of electricity and the meter. With this capability, providers have access to real-time information on the electricity consumption of each customer. Smart metering technology will be the foundation of any smart grid strategy, as it provides the platform for offering digital-era services to consumers.

With detailed, real-time information, providers of electricity can offer customers new, innovative rate forms commonly referred to as dynamic pricing. Rather than simply charging the same price for each unit of electricity that is consumed, dynamic pricing offers customers lower rates during those off-peak times when it is cheaper than average to provide electricity, and higher rates during peak times when it costs more to provide the electricity. With this information, customers have the opportunity to reduce their electricity bills by shifting consumption from higher-priced hours to lower priced hours, or through overall conservation. Dynamic pricing is currently being tested through a number of experimental pilots around the country and will become the default rate for most California customer classes by 2012.

Smart metering, when coupled with dynamic pricing, will also create a market for new smart grid consumer products. For example, programmable communicating thermostats (PCTs) could be used to automatically modify the setpoint when the price of electricity exceeds a pre-specified threshold, thus lowering customer bills. These automating technologies aren't limited to heating and cooling. The concept of appliances that "listen" to the price of electricity and operate accordingly can be extended to a number of residential, commercial, and industrial end uses. This concept of transmitting "prices to devices" was demonstrated in a pilot conducted by utilities in Washington State's Olympic Peninsula.

Another commercially available smart grid consumer product is the in-home information display (IHD). Whereas the smart meter provides realtime electricity consumption data to the provider of electricity, the IHD provides this information to the consumer. The IHD essentially acts as a speedometer for the customer's electricity consumption. It can provide information on historical consumption patterns and even give recommendations for easy ways to consume electricity more efficiently to reduce costs. By increasing customer awareness of the relationship between the amount of electricity they consume and the cost of consuming it, IHDs have been shown to have an overall conservation effect. IHDs can take many forms, from Internet websites to simple electrical socket plug-ins.

The smart grid components that have been described thus far AMI, dynamic pricing, automating technologies, and inhome displays are all commercially available and could be deployed on a large scale in the short term (for example, many utilities have announced AMI deployment plans for the next five to seven years which, if implemented, would result in a 40 percent market penetration of advanced metering technology). However, longer-term visions of the smart grid go well beyond these components to include technologies that are still early in their development stages. These include smart distribution networks, increased penetration of large-scale distributed generation and storage technologies, and plug-in hybrid electric vehicles (PHEVs).

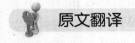 原文翻译

第二节 智能电网的定义

"智能电网"对不同的人来说具有不同的意义。但是，在现今多种定义中有一些共识存在。在基本层面上，智能电网将作为信息技术的支撑，使得当今电网无法支持的新技术得以广泛普及。这些新技术包括计量、传输、分配和电力存储技术方面的尖端技术，以及为消费者和电力供应商提供新的信息和灵活性。最后，应用这些信息将会改善供给消费者的产品和服务，形成更有效率的电力消耗和供应。

在所有智能电网定义中，最核心的部分是先进的测量设备（AMI）。这指的是智能仪表和伴随着的能够实现电能供应者和仪表双向交流的通信网络。有了这种能力，供应者能知道每一个消费者实时的用电情况。智能仪表技术将成为所有智能电网决策的基础，因为它可以给消费者提供数字时代的服务。

有了详细的实时信息，在创新的费率形式下，电能供应者能给消费者提供新的动态定价。与简单的将消耗的电能以每单元进行相同的计费相比，动态定价为客户在非高峰时间提供比平均电价更低的费率；同时在高峰时间因此时所耗成本更高，此时费率较高。有了这些信息，消费者能够有机会通过将高价时段的消费转移到低价时段的消费，或者通过全部保存，从而减少他们的电费。动态定价在全国范围内通过一系列测试，并且截至2012年，成为了大部分加利福尼亚消费者的计费方式。

和动态电价配合使用的智能电表也将创造一个新的智能电网消费产品市场。例如，可编程通信恒温器（PCT）可以用于在电价超过预定阈值时自动修改设定点，从而降低客户消费。这些自动控制技术不仅局限于制热或者制冷。依据电价运行的概念能被扩展到大量的居民，商业的和工业终端总应用中。在华盛顿州奥林匹克半岛的公用事业公司进行的试点证明了这种将"电价传递到设备"的概念的正确性。

另一个用于智能电网消费者的商业产品是室内信息显示器（IHD）。智能电表提供了实时的电能消耗数据给电能供应商，而室内信息显示器把这些数据提供给消费者。室内信息显示器本质上担当着消费者消耗电能的速度计。它能够为消费者提供历史消费方式信息，甚至能够提供简单的方法来使电能消耗更有效率，从而降低电费。通过提高客户对电量和消费成本之间关系的认识，IHDs已被证明具有整体的保护效果。IHDs可以采取多种形式，从互联网网站到简单的电插座插件。

到目前为止，智能电网元件已经被描述为先进的 AMI、动态电价、自动控制技术和室内显示的智能电网的组成部分都可用于商业化，而且可以在短期内大规模的进行调度（例如，许多公用设施已经公布了下一个五到七年间的先进测量设施调度计划，这些计划如果实施完成了，将会导致先进计量技术市场达到40%的渗透率），不管怎样，智能电网的长期展望将会比这些组成部分更好，包括在它们发展阶段之前出现的技术。这些包括智能配电网、存储技术和混合动力车。

New Words and Expressions

infrastructure	n. 基础设施
setpoint	n. 选点
AMI	先进的计量基础设施
electricity storage	电力存储
two-way communication	双向通电
electricity bill	电费
dynamic pricing	动态定价
socket	n. 插座
	vt. 给……配插座

 Notes

1. At a basic level, the smart grid will serve as the information technology backbone that enables widespread penetration of new technologies that today's electrical grid cannot support.

在基础层面上，智能电网将担任信息技术的骨干作用，这个作用能使在当今的输电网络无法支持的新技术大范围渗透。

2. These new technologies include cutting-edge advancements in metering, transmission, distribution, and electricity storage technology, as well as providing new information and flexibility to both consumers and providers of electricity.

这些新技术包括在计量、传输、分配、电能储存和对消费者以及电力生产者之间提供新信息及灵活性这些尖端技术。

3. Rather than simply charging the same price for each unit of electricity that is consumed, dynamic pricing offers customers lower rates during those off-peak times when it is cheaper than average to provide electricity, and higher rates during peak times when it costs more to provide the electricity.

并不是简单地将消耗的电能以每单元进行相同的计费，动态定价给消费者在那些当提供电能比平均费率低时的非高峰时期提供较低的费率，而当提供电能将花费更多时的高峰期提供较高的费率。

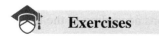

Ⅰ. Choose the best answer into the blank.

1. In estimating the potential environmental impacts of the smart grid, it is critical to account for regional system_____.

    A. motivate   B. dynamic   C. changed   D. varied

2. At the core of all smart grid definitions is advanced_____.

A. metering infrastructure   B. smart grid
C. dynamic pricing   D. PCT

3. Whereas the smart meter provides _____ electricity consumption data to the provider of electricity.

A. virtual   B. visual   C. realtime   D. recent

4. Ultimately, access to this information will improve the _____ that are offered to consumers.

A. products and services   B. quantity and services
C. effect and products   D. products and effect

5. Smart metering technology will be the foundation of any smart grid strategy, as it provides the platform for offering _____ services to consumers.

A. modernism   B. modernize   C. digital   D. digital-era

II. Answer the following questions according to the text.

1. What kind of role does smart grid serve as?
2. What is the advantages of AMI?
3. How do people use smart metering accompany with dynamic pricing?
4. How does IHD works?

III. Translate the following into Chinese.

The smart grid components that have been described thus far AMI, dynamic pricing, automating technologies, and inhome displays are all commercially available and could be deployed on a large scale in the short term (for example, many utilities have announced AMI deployment plans for the next five to seven years which, if implemented, would result in a 40 percent market penetration of advanced metering technology). However, longer-term visions of the smart grid go well beyond these components to include technologies that are still early in their development stages. These include smart distribution networks, increased penetration of large-scale distributed generation and storage technologies, and plug-in hybrid electric vehicles (PHEVs).

Keys

I. Choose the best answer into the blank.
1. B   2. A   3. C   4. A   5. D

II. Answer the following questions according to the text.

1. At a basic level, the smart grid will serve as the information technology backbone that enables widespread penetration of new technologies that today's electrical grid cannot support.

2. At the core of all smart grid definitions is advanced metering infrastructure (AMI). This refers to smart meters and an accompanying communications network that allows for two-way communication between the provider of electricity and the meter. With this capability, providers have access to real-time information on the electricity consumption of each customer.

Smart metering technology will be the foundation of any smart grid strategy, as it provides the platform for offering digital-era services to consumers.

3. For example, programmable communicating thermostats (PCTs) could be used to automatically modify the setpoint when the price of electricity exceeds a pre-specified threshold, thus lowering customer bills. These automating technologies aren't limited to heating and cooling.

III. Translate the following into Chinese.

到目前为止,已经被描述为先进的测量设施,动态电价、自动控制技术和室内显示的智能电网的组成部分都是可用于商业化,而且可以在短期内大规模的进行调度(例如,许多公用设施已经公布了下一个五到七年间的先进测量设施调度计划,这些计划如果实施完成了,将会导致先进计量技术市场达到40%的渗透率),不管怎样,智能电网的长期展望将会比这些组成部分更好,包括在它发展阶段之前出现的技术。这些包括智能配电网、存储技术和混合动力车。

翻译技巧之名词性从句(一)

名词在句子中的成分有可能是:主语、宾语、表语和同位语。相应地,名词性从句有四类:主语从句、宾语从句、同位语从句和表语从句。这四类名词性从句翻译过程中一般都遵循一个原则,如果翻译成汉语句子比较简单,不会造成主句的失衡现象,就可以翻译成从句本来应该承担的成分,进行顺译;但是,如果从句结构较为复杂,而主句较为简单,则适宜把从句单独成句翻译,主句中使用代词指代这个从句。下面我们来具体看一下翻译方法:

构成主语从句的方式有下列两种:

(1) 关联词或从属连词位于句首的从句+主句谓语+其他成分。

它们一般是译在句首,作为主从复合句的主语。这样的词有关联词 what,which,how,why,where,who,whatever,whoever,whenever,wherever 及从属连词 that,whether,if。如:

例题 1:What he told me was only half-truth.

分析:其中关联词可译"…的",放在后面。

参考译文:他告诉我的只是些半真半假的东西而已。

例题 2:Whether an organism is a plant or an animal sometimes taxes the brain of a biologist.

分析:whether 可以译成"是否,是…还是",然后适当安排位置。

参考译文:一种生物究竟是植物还是动物,有时使生物学家颇伤脑筋。

(2) it+谓语+that(whether)引导的从句。

如果先译主句,可以顺译为无人称句。有时也可先译从句,再译主句。如果先译从句,便可以在主句前加译。如:

例题 3:It can be seen that precleaning alone would not reduce the total sulfur content of the four coals to levels anywhere near the standards.

分析:这样的句子可以译成无人称句,常用来表示事实、常理等。

参考译文：可以看出，这四种煤的总含硫量仅靠预先清洗将不能降低到完全接近标准规定的水平。

例题 4：It is a matter of common experience that bodies are lighter in water than they are in air.

分析：有时为了使译文成分完整，可以补充上泛指的主语（如人们…、大家…）。类似的结构还有：

it is（universally）known that… 大家都知道…

it is believed that… 人们都相信…

参考译文：物体在水中比在空气中轻，这是一种大家共有的经验。

Section 3　The Expanded Scenario

An analysis of the smart grid should not stop with those technologies that are commercially available today. As described previously, the smart grid will enable technology adoption that could lead to a transformational change in the power industry. These changes include adoption of a nation-wide smart distribution system, increased penetration of renewable resources, widespread deployment of large-scale electricity storage devices, and adoption of PHEVs.

A national strategy including these changes is not out of the question. Consider, for example, Google's recently published Clean Energy 2030 Plan, a national strategy to dramatically reduce US dependence on foreign oil and cut national CO_2 emissions in half. This strategy includes deployment of a smart grid that will enable 90 percent penetration of PHEVs by 2030 and add over 700GW of renewable generation during this timeframe.

Similarly, President Obama's "New Energy for America" plan identifies the opportunities in distributed storage and renewable energy that the smart grid will provide. Specifically, "Barack Obama and Joe Biden will pursue a major investment in our national utility grid using smart metering, distributed storage, and other advanced technologies to accommodate 21st century energy requirements:It greatly improved electric grid reliability and security, a tremendous increase in renewable generation and greater customer choice and energy affordability."

Further evidence that an expanded smart grid strategy could be on the horizon is in former Vice President Al Gore's recent announcement that $400 billion should be invested in the national electricity infrastructure over the next 10 years. Among its many benefits, Gore cites the ability of this "modern grid" to move renewable electricity from the rural places where it is generated to the urban locations where it is mostly used. He also identifies the ability of such a system to accommodate PHEVs.

Such support from both the public and private sectors for a broadly expanded national upgrade of the electricity infrastructure cannot be ignored. These more aggressive smart grid strategies are represented in the Expanded Scenario. The Expanded Scenario includes the impacts of a higher penetration of renewables that would be enabled by the smart grid, and the effect of an accompanying increase in distributed storage that would be used to "back up" these sources of renewable energy. In addition, the Expanded Scenario includes the commercially available smart grid technologies described in the Conservative Scenario, and assumes the same impacts.

As described earlier, an increase in the penetration of renewable energy is enabled by the smart grid in two ways:

(1) the increased flexibility of the smart distribution system allows for greater penetration of distributed renewable resources such as residential solar panels;

(2) an expanded T&D system provides access to rich sources of previously untapped renewable energy such as wind power in more rural areas. Both could significantly increase the amount of renewable generation. However, determining exactly how significant the increase in renewables penetration will be is difficult.

As a starting point for this analysis it was assumed that by 2030 states would be able to achieve renewable penetration levels that are twice as high as the requirements of the existing renewable portfolio standards (RPS). This assumption accomplishes two objectives. First, it remains consistent with existing regional variations in political willingness to pursue renewable energy. States with a higher RPS are also more likely to aggressively pursue renewable energy in a smart grid framework. Second, to the extent that the RPS reflects the regional availability of renewable resources, the assumption accounts for regional differences in the level of renewable resources that can be accessed. This assumption to double the amount of generation from renewable sources is supported by a recent NREL study, which found that the battery storage capability of a large-scale deployment of PHEVs would result in an increase in wind generation of 105 percent.

On the surface, doubling the existing RPS appears to be a very aggressive assumption. With this assumption, 19 percent of US electricity generation would come from renewables in 2030 (today we get roughly 9 percent, including large-scale hydro). However, larger penetrations of renewable energy are being considered in many well-publicized energy plans. For example, T. Boone Pickens' national energy plan calls for 20 percent of US electricity generation to come from wind power alone in the next 10 years. The New Apollo program, developed by the Apollo Alliance, proposes a goal of 25 percent renewable energy by 2025. The previously described Google energy plan assumes that around 60 percent of the nation's electricity will come from renewable energy by 2030. And the Repower America project, inspired by Gore's call for 100 percent of US electricity to come from carbon-free sources in the next 10 years, has suggested that 75 percent of the nation's electricity come from renewables.

第三节 扩 展 方 案

现在对智能电网的分析不应停留于商业上的可用技术。如前所述，智能电网将通过采用相关技术，实现电力工业的转型变革。这些变革包括全国范围内智能配电系统的使用、可再生能源普及率的持续增长、大型电力储能装置的广泛布设和插电式混合动力汽车的推广。

一项包含这些变革的国家战略并不是天方夜谭。例如，谷歌最近公布不了一项国家战略——2030年清洁能源计划，旨在迅速减少美国对外国石油的依赖，以及将国家的二氧化碳排放量降低一半。在该战略中部署了一个智能电网，到2030年其将使插电式混合动力汽车的普及率达到90%，同时，在该计划期间内增加超过700GW的可再生能源发电量。

同样的，奥巴马总统的"美国新能源"计划意识到，智能电网将在分布式存储、可再生能源方面具有优势与机遇。具体地说，"巴拉克·奥巴马和乔·拜登将在国家公用电网中进行重大投资，通过使用智能电能表、分布式存储和其他先进技术满足21世纪的能源需求，大幅提高电网的可靠性和安全性，同时实现可再生能源发电量的极大增长、客户选择面的拓宽和能源可供应力的增强。"

一项扩展的智能电网战略即将浮出水面，这也在前副总统阿尔戈尔最近的声明中得到了进一步验证。他在该声明中表示，在未来10年内，将有4000亿美元被用于国家电力基础设施的投资建设上。在智能电网带来的诸多益处中，戈尔引用了"现代电网"将乡村地区生产的可再生电能输送至耗电量大的城市地区的能力，同时他也确认了这样一个系统能够服务于插电式混合动力汽车的能力。

不容忽视的一点是，全国范围内电力基础设施的全面扩张升级得到了公共及私营部门的支持。这些较为激进的智能电网战略在扩展方案中有所体现。扩展方案包括智能电网可实现的较高再生能源普及率的影响，以及随之提升用于"备份"这些可再生能源的分布式存储的效果。除此之外，扩展方案还包括保守方案中描述的商业可用智能电网技术，并具有相同的影响。

如前所述，智能电网实现可再生能源普及率增长的方式有两种：

（1）智能配电系统能够更加灵活地吸纳分布式可再生能源，比如居民住宅的太阳能；

（2）扩展的输配电系统可以将以前未开发的可再生能源，比如偏远乡村地区的风能接入电网。上述两种方式都能极大地提升可再生能源发电的总量。但是，很难知道具体能够提升多少可再生能源发电量。

作为智能电网分析的开端，设想到2030年，国家将使可再生能源的覆盖水平达到可再生能源配额（RPS）要求的两倍。该设想将实现两个目标。第一，现有寻求可再生能源的政治意愿存在区域差异，这与它是保持一致的。拥有较高可再生能源配额的州也更乐意在智能电网网架下积极主动寻求可再生能源。第二，在某种程度上讲，可再生能源配额反映了区域的可再生能源可利用量，所以可再生能源的拥有量会因区域的不同而存在差别，从而可以得到这个设想。最近，国家可再生能源实验室的研究佐证了可再生资源发电量翻倍的设想，发现大规模部署的插电式混合动力汽车的电池容量，将使风力发电的增长量达到105%。

从表面上看，实现现有可再生能源配额的翻倍貌似是一个非常激进的设想。按照该设想，在2030年19%的美国电力生产将来自于可再生能源（包括大型水电在内，如今可再生能源发电大概占9%）。但是，在许多众所周知的能源计划中都提到了可再生能源的大范围普及。例如，特布恩·皮肯斯的国家能源计划提倡在下一个10年内，使20%的美国电力生产完全来自于风能；阿波罗联盟主导的《新阿波罗计划》提出一个目标，到 2025 年实现 25%的电能来自于可再生能源；之前提到的谷歌能源计划设想到 2030 年，有 60%的国家电力来自于可再生能源；戈尔提倡在下一个 10 年内，100%的美国电力来自于无碳能源，而《美国新动力》项目正是受其启发，建议 75%的美国电力取自于可再生能源。

New Words and Expressions

transformational	*adj.* 转换的，转换生成的
PHEV	室内信息显示器
distribution system	分配制度；配电系统

Notes

1. These changes include adoption of a nation-wide smart distribution system, increased penetration of renewable resources, widespread deployment of large-scale electricity storage devices, and adoption of PHEVs.

这些变革包括全国范围内智能配电系统的使用、可再生能源普及率的持续增长、大型电力储能装置的广泛布设和插电式混合动力汽车的推广。

2. Specifically, "Barack Obama and Joe Biden will pursue a major investment in our national utility grid using smart metering, distributed storage, and other advanced technologies to accommodate 21st century energy requirements:It greatly improved electric grid reliability and security, a tremendous increase in renewable generation and greater customer choice and energy affordability."

具体地说，"巴拉克·奥巴马和乔·拜登将在国家公用电网中进行重大投资，通过使用智能电表、分布式存储和其他先进技术满足 21 世纪的能源需求，大幅提高电网的可靠性和安全性，同时实现可再生能源发电量的极大增长、客户选择面的拓宽和能源可供应力的增强。"

3. And the Repower America project, inspired by Gore's call for 100 percent of US electricity to come from carbon-free sources in the next 10years, has suggested that 75 percent of the nation's electricity come from renewables.

戈尔提倡在下一个 10 年内，100%的美国电力来自于无碳能源，而《美国新动力》项目正是受其启发，建议 75%的美国电力取自于可再生能源。

Exercises

Ⅰ. Choose the best answer into the blank.

1. Which followings would lead to a transformational change in the power industry? ___
    A. smart grid   B. modernize arrangement

C. smart computer   D. solar energy

2. A national strategy to dramatically reduce US dependence on foreign oil and cut national ___ emissions in half.
   A. CO   B. CO_2 C. SO_2 D. NO_2

3. Gore cites the ability of this "modern grid" to move renewable electricity from the ____ places where it is generated to the ____ locations where it is mostly used.
   A. urban;rural   B. urban;outside
   C. rural;urban   D. rural;outside

4. As a starting point for this analysis it was assumed that by 2030 states would be able to achieve renewable ____ levels that are twice as high as the ____ of the existing renewable portfolio standards (RPS).
   A. requirements;penetration   B. penetration;acquirement
   C. acquirement;penetration   D. penetration;requirements

5. For example, T. Boone Pickens' national energy plan calls for 20 percent of US electricity generation to come from ____ power alone in the next 10years.
   A. wind   B. water   C. human-made   D. coal

II. Answer the following questions according to the text.

1. Smart grid will cause what kind of changes in power industry?
2. what are the two ways does smart grid enable the penetration of renewable energy?
3. what purposes will be accomplished by this assumption?

III. Translate the following into Chinese.

As described earlier, an increase in the penetration of renewable energy is enabled by the smart grid in two ways:

(1) the increased flexibility of the smart distribution system allows for greater penetration of distributed renewable resources such as residential solar panels;

(2) an expanded T&D system provides access to rich sources of previously untapped renewable energy such as wind power in more rural areas. Both could significantly increase the amount of renewable generation. However, determining exactly how significant the increase in renewables penetration will be is difficult.

Keys

I. Choose the best answer into the blank.
1. A   2. B   3. C   4. D   5. A

II. Answer the following questions according to the text.

1. As described previously, the smart grid will enable technology adoption that could lead to a transformational change in the power industry. These changes include adoption of a nation-wide smart distribution system, increased penetration of renewable resources, widespread deployment of large-scale electricity storage devices, and adoption of PHEVs.

2. As described previously, the smart grid will enable technology adoption that could lead to a transformational change in the power industry. These changes include adoption of a nation-wide smart distribution system, increased penetration of renewable resources, widespread deployment of large-scale electricity storage devices, and adoption of PHEVs.

3. First, it remains consistent with existing regional variations in political willingness to pursue renewable energy. States with a higher RPS are also more likely to aggressively pursue renewable energy in a smart grid framework. Second, to the extent that the RPS reflects the regional availability of renewable resources, the assumption accounts for regional differences in the level of renewable resources that can be accessed.

Ⅲ. Translate the following into Chinese.

如前所述，智能电网实现可再生能源普及率增长的方式有两种：

（1）智能配电系统能够更加灵活地吸纳分布式可再生能源，比如居民住宅的太阳能；

（2）扩展的输配电系统可以将以前未开发的可再生能源，比如偏远乡村地区的风能接入电网。上述两种方式都能极大地提升可再生能源发电的总量。但是，很难知道具体能够提升多少可再生能源发电量。

翻译技巧之名词性从句（二）

构成表语从句的方式有下列两种：

（1）表语从句是位于主句的联系动词后面、充当主句主语的表语从句，它也是由 that，what，why，how，when，where，whether 等连词和关联词引导的。一般来讲，可以先译主句，后译从句。如：

例题 5：The result of invention of steam engine was that human power was replaced by mechanical power.

分析：先译后面的主句，后译前面的从句。

参考译文：蒸汽机发明的结果是机械力代替了人力。

（2）几种常见句型，以下逐一介绍它们的译法。

★在 that（this）is why……句型中，如果选择先译主句，后译从句，可以译成这就是为什么……，这就是为什么……的原因，这就是……的缘故等。如果选择先译从句，再译主句，一般可以译为……原因就在这里，……理由就在这里等。如：

例题 6：That is why practice is the criterion of truth and why the standard of practice should be first and fundamental in the theory of knowledge。

分析：原句表语很长，所以此处采用了紧缩原则。

参考译文：所谓实践是真理的标准，所谓实践的标准，应该是认识论的首先的和基本的观点，理由就在这个地方。

★在 this（it）is because……句型中，一般先译主句，再译从句，译成是因为……，这是因为……的缘故，这是由于……的缘故。如：

例题 7：This is because the direct current flows in a wire always in one direction.

参考译文：这是由于直流电在导线中总沿着一个方向流动的缘故。

★在 this is what……句型中,如果先译主句,后译从句,通常译为这就是……的内容,这就是……的含义等。如果先译从句,后译主句,通常译为……就是这个道理,……就是这个意思等。如:

例题 8: This is what we have discussed in this article.

分析:如表语不长,则可译成这就是……的内容等。

参考译文:这就是我们在本文中所讨论的内容。

Chapter 10   The Introduction of SGCC and CSG

Section 1   Brief Introduction of State Grid Corporation of China (SGCC)

Established on the basis of a sum of enterprises and institutions formerly owned by State Power Corporation of China, State Grid Corporation of China (short for SG) is a special large-sized enterprise approved by the State Council to operate the business of power transmission, transformation, distribution and other assets of power grid. As a pilot state-holding and authorized investing Corporation by the State Council, whose leading group of is under the administration and management of the Central Government directly, SG operates in such a manner that the president as the legal person, is responsible for all major activities.

At present, there are 728 thousand staffs totally working for SG, among which 522 thousand staffs (71.7% of the total) are engaged in the field of industry, 157 thousand staffs (21.6% of the total) are engaged in the field of architecture and 14.6 thousand staffs (2% of the total) are working in the research and design institutes.

1. Major Responsibilities

To operate the Corporation in accordance with state laws, regulations and industrial policies, corresponding state macro control and sector supervision principles with the focus on market demand.

To operate, manage and supervise the assets invested by state and owned by the Corporation, undertaking responsibilities to preserve and add value on the assets.

To formulate and conduct development strategy, mid and long term plan, annual plan and significant operation and managing decisions of the Corporation following mid and long term plan of national economy, industrial policy of state, development plan of power industry and market demand; to assist formulation of national grid development plan and make proposal on development plan on national power industry.

To participate, construct and operate related trans-regional connection projects; to be responsible for construction and management of transmission and transforming project connected to the Three Gorges Power Station.

To be responsible for power transaction and dispatching among the power grids within service area, and achieve safe, efficient operation through coordination of daily business among grids.

To optimize resources allocation and production elements by conducting important investment with efficient output in accordance with state laws, regulations and related policies.

To keep stability of the Corporation and society in the process of deepening reform and structural adjustment by transforming operation mechanism, strengthening internal management, simplifying organization and re-employment of laid-offs.

To reinforce corporate culture and corporate value of the Corporation and make integrated management on intangible assets regarding the name, brand and credibility of the Corporation.

To undertake other matters entrusted by the State Council.

2. Business scope

To operate all state-owned assets owned by the Corporation in accordance with state laws.

To purchase and sell electricity, to dispatch and trade power among grids within service area.

To invest, construct, operate and manage trans-regional transmission and transformation projects.

To carry out investment and financing business both at home and abroad approved by related authorities according to state regulations.

To conduct foreign trade, international cooperation, overseas project contracting and export of labor service.

To engage in scientific research, technical development, power dispatching and communication as well as consulting service in power supply business.

To operate other business approved by state.

3. Major authorities

To re-invest and adjust structure of state-owned capital according to "Corporation Law" by utilizing part of profits from state-owned assets with the precondition to ensure the legal rights and interests of its self development.

To make decision on the restructure, transfer, lease of state-owned assets and merger & acquisition business of wholly owned subsidiaries. The same matter of above mentioned for stock holding or sharing enterprises is decided or suggested by the Corporation through legal procedure.

To decide on investment. To operate and manage projects invested by the Corporation and related enterprises according to related regulations.

To examine and approve foreign affaires related to the Corporation, to engage in foreign trade, international investment and financing, external guarantee, overseas project contracting and export of labor service.

To set up corporate governance system and institutional structure. To decide the mode of operation and distribution, significant corporate issues as well as the merger, partition and dissolving of related enterprises.

To appoint, remove and manage the executives of wholly owned subsidiaries and headquarter of the Corporation according to jurisdiction and procedure of personnel appointment. To delegate or replace the shareholder representatives of stockholding or sharing enterprises, and recommend the board of directors and members of supervision board according to legal procedure and capital proportion.

To integrate management on the foreign affairs related to the Corporation and relevant enterprises. To approve overseas trips of senior executives of the Corporation.

Other authorities vested by the State Council and related government agencies.

4. Development mission

Mission: To build a modernized corporation with strong power grid, excellent assets, service and performance.

Development: To promote the trans-region and trans-valley complementation of thermal power and hydropower and optimize re-allocation of resources in larger scope, to speed up planning of national backbone network structure, optimized regional interconnection and corresponding voltage level.

Commitment: To adhere to sincere service and win-win development. To be committed on sincere, professional and first class services to power generation companies, power customers and make contribution to development of the society and the country.

Corporate governance: To be a leading corporation by focusing on development, management and staffing. Great importance will be attached to strengthening internal management, synergizing managerial mechanism and enhancing incentive mechanism. To improve managerial efficiency by optimized and standard administrative process as well as sufficient information exchange. To evaluate the Corporation by performance target and result achieved.

第一节 国家电网公司简介（SGCC）

国家电网公司（简称SG）是从原国家电力公司旗下的一系列公司和机构中剥离出来，并经国务院批准的特殊大型企业，负责运行和经营电力传输、转换、分配及电网其他资产等业务。作为经过国务院同意进行国家授权投资的机构和国家控股公司的试点单位，公司领导团队受中央政府直接管理，公司实行总经理负责制，总经理是公司的法定代表人。

目前，国家电网有72.8万公司员工，其中52.2万名员工（占总数的71.7%）从事工业领域，15.7万员工（占总数的21.6%）从事建筑领域，还有1.46万员工（占总数的2%）工作于研究所和设计院。

1. 企业职责

执行国家法律、法规和产业政策，在国家宏观调控和行业监管下，以市场需求为导向，依法自主经营。

对有关企业中由国家投资形成并由国家电网公司拥有的国有资产依法经营、管理和监督，并相应承担保值增值责任。

根据国民经济中长期发展规划、国家产业政策、电力工业发展规划和市场需求，制订并组织实施国家电网公司的发展战略、中长期发展规划、年度计划和公司重大生产经营决策，协助制订全国电网发展规划，提出全国电力工业发展规划的建议。

参与、建设和经营相关的跨区域输变电和联网工程；负责连接至三峡电厂的输变电工程的建造与管理。

负责所辖各区域电网之间的电力交易和调度，处理区域电网公司日常生产中的网间协调问题，实现安全、高效运行。

根据国家法律、法规和有关政策，依据投入产出效果组织实施重大投资活动，优化配置生产要素。

通过转换企业经营机制、强化内部管理、精简机构和富余人员分流与再就业工作，深化企业改革，加快结构调整，维护企业和社会稳定。

加强企业文化建设和有关企业价值工作，统一管理国家电网公司的名称、商标和商标信誉等无形资产。

承担国务院委托的其他工作。

2. 业务范围

依法经营国家电网公司拥有的全部国有资产。

从事电力购销业务，负责所辖各区域电网之间的电力交易和调度。

参与投资、建设、经营和管理相关的跨区域输变电和联网工程。

根据国家有关规定，经相关部门批准，从事国内外投融资业务。

开展外贸流通经营、国际合作、对外工程承包和对外劳务合作等业务。

从事与电力供应有关的科学研究、技术开发、电力生产调度信息通信、咨询服务等业务。

经营国家批准或者允许的其他业务。

3. 主要权力

在保证自身发展和合法权益的前提下，利用国有资产的部分利润，根据"公司法"进行再投资和调整国有资本结构。

决定全资子公司的重组、转让、租赁和合并收购业务。如上所述，公司通过法律程序决定或建议持有或分配股份的企业。

做投资决定。按照相关规定，经营和管理公司及相关企业的投资项目。

核对审批公司相关跨国事务，从事对外贸易、国际投资和融资、对外担保、海外工程承包和劳务出口。

建立公司治理系统和制度结构。决定运营、分配模式，以及相关企业的合并、划分和解体等重大企业问题。

根据人事部门的管辖权和程序任命、调动和管理全资子公司经理，以及设立公司总部。委派或更换持有股票或股份的企业股东代表，根据法律程序和资金比例推荐董事会和监事会成员。

集成管理总公司及相关企业的外交事务。任命海外公司的高管。

承担国务院及有关部门委托的其他工作。

4. 发展任务

任务：建立拥有可靠电网、雄厚资金、优质服务和良好业绩的现代化公司。

发展：促进跨区域、跨流域火力发电和水力发电的互补，大范围优化资源再分配。加快国家骨干电网结构规划，优化区域互联匹配电压等级。

承诺：坚持真诚服务和双赢发展。致力于真诚、专业、一流的服务发电公司和电力客户，为社会和国家发展做出贡献。

企业治理：成为专注开发、管理和人员配置的领先公司。重视加强内部管理、增效管理机制、强化激励机制。通过优化标准的行政程序与充分的信息交换改善管理效率。评估公司的业绩目标和取得成就。

New Words and Expressions

dispatch	*n.* 调度
trans-valley	跨流域
thermalpower	*n.* 火力发电机
national backbone network structure	国家骨干电网
incentive	*n.* 动机
	adj. 激励的，刺激的

Notes

1. Established on the basis of a sum of enterprises and institutions formerly owned by State Power Corporation of China, State Grid Corporation of China (short for SG) is a special large-sized enterprise approved by the State Council to operate the business of power transmission, transformation, distribution and other assets of power grid.

国家电网公司（简称 SG）是从原国家电力公司旗下的一系列公司和机构中剥离出来，并经国务院批准的特殊大型企业，负责运行和经营电力传输、转换、分配及电网其他资产等业务。

2. To appoint, remove and manage the executives of wholly owned subsidiaries and headquarter of the Corporation according to jurisdiction and procedure of personnel appointment. To delegate or replace the shareholder representatives of stockholding or sharing enterprises, and recommend the board of directors and members of supervision board according to legal procedure and capital proportion.

根据人事部门的管辖权和程序任命、调动和管理全资子公司经理，以及设立公司总部。委派或更换持有股票或股份的企业股东代表，根据法律程序和资金比例推荐董事会和监事会成员。

Exercises

Ⅰ. Choose the best answer into the blank.

1. State Grid Corporation of China is a special large-sized enterprise, which achieved the approval of ____.

A. State Council B. State Government
C. Provincial Government D. County Government

2. Which of the follow is the intangible assets of State Grid Corporation of China? ___
    A. transmission lines B. brand
    C. power equipment D. customer

3. State Grid Corporation of China can re-invest and adjust structure of state-owned capital according to "Corporation Law" by utilizing part of ____ from state-owned assets with the precondition to ensure the legal rights and interests of its self development.
    A. assets B. tax C. profits D. fixed assets

4. State Grid Corporation of China promote the complementation of ____ and hydropower.
    A. thermal power B. wind power C. solar power D. tidal power

5. State Grid Corporation of China emphasize ____ management.
    A. external B. synergy C. staff D. internal

Ⅱ. Answer the following questions according to the text.

1. Who owned the administration and management of State Grid Corporation of China?

2. What should be down if the State Grid Corporation of China operate other business?

3. What is the mission of the State Grid Corporation of China?

Ⅲ. Translate the following into Chinese.

The State Grid Corporation of China(Grid State), referred to as the State Grid, the national network, was established in December 29, 2002, is the State Council approved the state authorized investment institutions and state holding company's pilot units.The company as a state-owned key enterprises related to national energy security and national economy, the construction and operation of power grid as the core business, undertakes the basic mission of power supply security more secure, more economical, clean and sustainable business area, covering 26 provinces (autonomous regions and municipalities), covering the land area 88%, the power supply more than 1 billion 100 million people, the total number of employees more than 1 million 860 thousand people.Companies in Brazil, Philippines, Portugal, Australia and other countries and regions to carry out business. In 2016, the company ranked "fortune" the world 500 strong second , is the world's largest public utility enterprises. General manager is responsible for the implementation of the company, the general manager is the legal representative of the company.

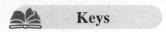

Keys

Ⅰ. Choose the best answer into the blank.
1. A 2. B 3. C 4. A 5. D

Ⅱ. Answer the following questions according to the text.

1. the Central Government.

2. The State Grid Corporation of China should gain the approval of state.

3. To build a modernized corporation with strong power grid, excellent assets, service and performance.

III. Translate the following into Chinese.

国家电网公司（State Grid），又称为国家电网、国网，成立于2002年12月29日，是经过国务院同意进行国家授权投资的机构和国家控股公司的试点单位。公司作为关系国家能源安全和国民经济命脉的国有重要骨干企业，以建设和运营电网为核心业务，承担着保障更安全、更经济、更清洁、可持续的电力供应的基本使命，经营区域覆盖全国26个省（自治区、直辖市），覆盖88%的国土面积，供电人口超过11亿人，公司员工总量超过186万人。公司在菲律宾、巴西、葡萄牙、澳大利亚等国家和地区开展业务。2016年，公司位列"财富"世界500强第2名，是全球最大的公用事业企业。公司实行总经理负责制，总经理是公司的法定代表人。

翻译技巧之名词性从句（三）

构成宾语从句的方式有下列两种：

（1）用that，what，how，when，which，why，whether。if等引起的宾语从句，翻译成汉语的时候，一般不需要改变它在原句中的顺序。

I told him that because of the last condition, I'd have to turn it down.
我告诉他，由于那最后一个条件，我只得谢绝。

Can you hear what I say?
你听得到我所讲的吗？

I don't know that he swam across the river.
我不知道他游过了那条河。

I don't know how he swam across the river.
我不知道他是怎么游过那条河的。

He has informed me when they are to discuss my proposal.
他已经通知我他们将在什么时候谈论我的建议。

有时可加"说"字，再接下去翻译英语原文宾语从句的内容。

Smith replied that he was sorry.
斯密斯回答说，他感到遗憾。

He would remind people again that it was decided not only by himself but by lots of others.
他再三提醒大家说，决定这件事的不只是他一个人，还有其他许多人。

（2）用it作形式宾语的句子，在翻译的时候，that所引导的宾语从句一般可按英语原文顺序翻译；it有时候可以不用翻译。

I made it clear to them that they must hand in their papers before 10 o'clock in the morning.
我向他们讲清楚了的，他们必须在上午十时前交卷。（it没有翻译）

I heard it said that he had gone abroad.

听说他已经出国了。(it 没有翻译)

但有时候，也可以在译文中将 that 引导的宾语从句提前到句子最前面翻译。

I regard it as an honor that I am chosen to attend the meeting.

我被选参加会议，感到光荣。(it 没有翻译)

We consider it absolutely necessary that we should open our door to the outside world.

打开国门，实行开放，我们认为这是绝对必要的。(it 翻译为"这")。

Section 2 SGCC Profit and Output on The Rise

The nation's largest electricity grid builder, the State Grid Corp of China (SGCC), made a pre-tax profit of 29.67 billion yuan (US $3.65 billion) in the first half of this year, as China's demand for power continues to soar.

Profits were up 34.5 percent up compared with the same period last year, Wang Min, spokesman of SGCC said at a press conference yesterday in Beijing.

The company collected sales revenue of 327.1 billion yuan (US $40 billion) from January to June, a year-on-year increase of 19.9 percent, achieved by selling 13.8 percent more electricity at 696.5 billion kilowatt-hours (kWh).

"Improved management, reduced costs, an intensified effort to expand the nation's power transmission lines, as well as China's surging demand for power, are behind the improvement of the company's performance in the first six months of the year," Wang yesterday said.

Wang Yonggan, secretary-general of the China Electricity Council, earlier predicted China's total electricity consumption for this year to increase 13.5 percent year on year to 2456 billion kWh.

And more transmission facilities will consequently be needed to transmit electricity from the resource-rich areas such as China's western provinces, to the economically developed coastal areas in the east.

By the end of June, SGCC had made fixed asset investments of 43.9 billion yuan (US $5.4 billion), 6.3 billion yuan (US $776 million) more than the same period last year.

Some 91.3 percent of the investment, a total of 40.1 billion yuan (US $4.9 billion) has been poured into construction in, and the upgrading of the electricity transmission network.

According to company sources, 6502 kilometres of transmission lines capable of handling over 220 kilovolts, had gone on stream during the six-month period, thereby benefiting most of the country's power consumers.

With the advanced transmission capacity, SGCC accordingly transmitted 16 percent more electricity across different regions and provinces year-on-year to reach 32 billion kWh in the first half.

SGCC recently completed a major power transmission project to link the northwest and central China grids, which marks a significant step forward in connecting the six provincial grids managed by SGCC and China Southern Grid (CSG).

The newly completed projects include a converter station based in Henan Province linking the central China and northwestern grids, as well as an upgrade to the existing Gezhouba-Nanqiao 500 kilovolt transmission line, which transmits electricity from Sichuan to Shanghai.

In an effort to further enhance the country's electricity grid capacity, SGCC have also initiated work into an ultra high-voltage transmission network. A major pilot project to transmit electricity from the coal-rich Shanxi Province to the central Hubei Province is well under way, said the company.

The feasibility study has been finished, and major approval procedures have been passed, with construction of the pilot infrastructure project to begin by the end of this year.

原文翻译

第二节 国家电网公司增长的利润及产量

随着中国电力需求的不断增长，全国最大的电网建造者——国家电网公司（SGCC）在今年上半年获得了 296.7 亿元（合 36.5 亿美元）的税前利润。

昨日在北京召开的新闻发布会上，公司发言人王敏说："公司利润较去年同期相比增长了 34.5%。"

从 1 月到 6 月，公司产品销售总收入达 3271 亿元（合 400 亿美元），同比增长 19.9%。该收入的增长是通过多销售 13.8% 的电量实现的，其中销售电量为 6965 亿 kWh。

王敏昨日表示："今年前六个月公司业绩的提升归结于改善管理、降低成本、我国输电线路的扩张以及我国电力需求的迅猛增加。"

中国电力企业联合会秘书长王勇敢早前预测，今年中国的总用电量将同比增加 13.5%，达到 2.456 万亿 kWh。

因此，需要更多的电力传输设备将资源丰富地区（如我国西部省份）的电能传送到经济发达的东部沿海地区。

到 6 月底，公司已经进行了 439 亿元（合 54 亿美元）固定资产投资，比去年同期增长了 63 亿元（合 7.76 亿美元）。

约 91.3% 的投资、共计 401 亿元（合 49 亿美元）已经投入到输电网络的建设和升级当中。

据国家电网公司说，长度为 6502km 且电压等级在 220kV 以上的输电线路在上半年六个月的周期内相继投运，使大部分电力用户受惠。

利用先进的输电技术，上半年国家电网公司跨越不同地区及省份的输电量同比增长 16%，达到 320 亿 kWh。

最近，国家电网公司完成了一项连接西北电网与华中电网的重大输电工程，标志着国家电网和南方电网朝着连接六个省域电网的目标迈出了重大一步。

最近完成的项目包括一个位于河南省的换流站，用于连接华中电网和西北电网，以及升级现有 500kV 葛洲坝至南桥的高压输电线路（该线路将电能从四川输送到上海）。

为了进一步增大我国电网的容量，公司也开始了一项建设特高压输电网的工作。公司表示，一项由煤矿资源丰富的山西省至湖北省的重大输电试点工程正在有序地进行。

伴随着今年年底试点基础设施建设项目的开始，该工程的可行性研究已经完成，同时主要审批程序也已经通过。

New Words and Expressions

converter station	换流站
China Electricity Council	中国电力企业联合会
transmission facility	传输设施；输电线

Notes

The company collected sales revenue of 327.1 billion yuan (US $40 billion) from January to June, a year-on-year increase of 19.9 per cent, achieved by selling 13.8 per cent more electricity at 696.5 billion kilowatt-hours (kWh).

从 1 月到 6 月，公司产品销售总收入达 3271 亿元（合 400 亿美元），同比增长 19.9%。该收入的增长是通过多销售 13.8%的电量实现的，其中销售电量为 6965 亿 kWh。

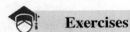
Exercises

Ⅰ. Choose the best answer into the blank.
1. Who is Wang Yonggan? ___
 A. spokesman of SGCC
 B. chairman
 C. secretary-general of the China Electricity Council
 D. provincial governor
2. By the end of June, SGCC had made fixed asset investments of ____ billion yuan.
 A. 6.3   B. 43.9   C. 29.67   D. 4.9
3. The newly completed projects include a ____ based in Henan Province.
 A. converter station   B. transformer substation
 C. rectifier   D. transformer
4. In an effort to further enhance the country's electricity grid capacity, SGCC have also initiated work into an ____ .
 A. ultra high-voltage transmission network   B. wireless transmission network
 C. superconductor transmission network   D. low current transmission network

Ⅱ. Answer the following questions according to the text.
1. What is the main reason of the improvement of the company's performance in the first

six months?

2. What is the main mean of linking the northwest and central China grids?

3. Why most power users benefit ?

Ⅲ. Translate the following into Chinese.

As China's demand for electricity continues to grow, The State Grid Corporation of China profits continue to increase. Wang Yonggan, the Secretary General of the China Electric Power Enterprise Association, said earlier that China's electricity power would be increased. Therefore,The State Grid Corporation of China increase the investment in power production, power transmission and energy dissipation in recent years,while the State Grid has begun to build and upgrade the transmission network. In order to further enhance the capacity of the State Grid, the company also launched an ultra high voltage transmission network work, some of the pilot infrastructure projects will be built at the end of the year.

Keys

Ⅰ. Choose the best answer into the blank.

1. C   2. B   3. B   4. A

Ⅱ. Answer the following questions according to the text.

1. Improved management, reduced costs, an intensified effort to expand the nation's power transmission lines, as well as China's surging demand for power.

2. It marks a significant step forward in connecting the six provincial grids managed by SGCC and China Southern Grid (CSG).

3. Because 6502 kilometres of transmission lines capable of handling over 220 kilovolts, had gone on stream during the six-month period.

Ⅲ. Translate the following into Chinese.

随着中国对电力需求的不断增长，国家电网的利润不断增加。中国电力企业联合会秘书长王勇敢早前表示，中国的电力将增长。因此，近些年国家电网在电能生产、电能传输和能量耗散等方面不断加大投资，与此同时，国家电网也不断开始建设与升级输电网络。为了进一步增加国家电网的容量，公司发起了一项超高压输电网络工作，一系列试点基础设施项目将在年底开始建设。

翻译技巧之名词性从句（四）

构成同位语从句的方式介绍如下：

同位语从句主要是用来对名词做进一步的解释，说明名词的具体内容。能接同位语从句的名词主要有：belief（相信）、fact（事实）、hope（希望）、idea（想法，观点）、doubt（怀疑）、news（新闻，消息）、rumor（传闻）、conclusion（结论）、evidence（证据）、suggestion（建议）、problem（问题）、order（命令）、answer（回答）、decision（决定）、discovery（发现）、explanation（解释）、information（消息）、knowledge（知识）、law（法律）、opinion

(意见，观点)、truth（真理，事实）、promise（承诺）、report（报告）、thought（思想）、statement（声明）、rule（规定）、possibility（可能）等。

（1）一般来说，同位语从句可以直接翻译在主句后面。

He expressed the hope that he would come over to visit China again.

他表示希望再到中国来访问。

There is a possibility that he is a spy.

有可能他是一个间谍。

（2）有时候在翻译同位语从句时，可以将其放在所修饰的名词前面，相当于前置的修饰语，但不一定使用定语的标志词"的"。这种情况下，同位语从句都是比较简单。

We know the fact that bodies possess weight.

我们都知道物体具有重量这一事实。

The rumor that he was arrested was unfounded.

他被逮捕的传闻是没有根据的。

Section 3　China Southern Power Grid

China Southern Power Grid Co., Ltd. (hereinafter referred to as CSG) was established on December 29th, 2002 in accordance with "The Power Sector De-regulatory Reform Program" promulgated by the State Council of China. CSG invests, constructs and operates power networks in Guangdong, Guangxi, Yunnan, Guizhou and Hainan provinces and regions. The service area is of 1 million square kilometers, with a population of 230million.

In 2009, CSG's electricity sale was 523.9TWh, a 6.2% increase than the previous year. The west-east transmission reached 115.6TWh, recording a 9.5% increase. The company revenues reached 313.6 billion RMB, increased by 9.8%. CSG spent 105.1 billion RMB in permanent assets investment. By the end of 2009, the total assets of CSG amounted to 442.5 billion RMB. The company ranks 185th in the 2009 Fortune Global 500.

In the service area of CSG, the total installed generation capacity reached 160GW, the 220kV-and-above transmission lines totaled at 76688km. Featuring long-distance, large capacity, ultra-high voltage and hybrid operation of AC and DC, CSG is operating one of the most sophisticated and technically advanced power grids in the world. In Jun. 2009, the completion of Hainan Interconnection Project ended the long history of "isolated" Hainan grid. Hainan grid was integrated into southern power grid by three 32km submarine cables. The submarine cables are the longest in Asia and the second longest in the world. The project is the first long-distance, large-volume 500kV cross-strait interconnection project in China. In Dec. 2009, CSG ±800kV DC transmission project started single pole operation. As the world's first ±800kV DC transmission project and China's UHV DC demonstration project, more than 60% of the project equipment is domestically manufactured, the single pole transmission capacity reached 2600MW. At present, CSG has thirteen EHV lines (eight AC lines, five DC lines) transmitting electricity from west to east, with total transmission capacity over 23GW.

Being adjacent to Hong Kong, Macau, and bordering Vietnam, the Laos and Cambodia, CSG service area enjoys unique geographical advantage in developing power cooperation with neighboring regions and countries. The company has been playing an active role within the Greater Mekong Sub-region(GMS) power cooperation framework, and is committed to providing quality supply and services to facilitate economic and social growth in the GMS countries. CSG supply electricity to Vietnam through seven interconnections(three 220kV and four 110kV), and the total export in 2009 was 4140GWh, a 26.6% increase. By the end of 2009, the 115kV Mengla(Yunnan, China) to Namo(Laos) interconnection was put into operation, starting electricity supply from CSG to Luang Nam Tha and Oudomxay in northern Laos. Besides, southern power grid is connected to Hong Kong grid through four 400kV circuits and several 132kV circuits; connected to Macau grid through three 220kV circuits and four 110kV circuits.

Speech of Board Chairman:

2008 is an extraordinary year. As the company experienced and overcame a series of daunting challenges and tests, we always puts the nation's and its people's interests at the top priority and firmly stick to our role as a central state-owned enterprise. We completely and promptly restored grid operation from damages made by the severe ice disaster; dedicated full efforts to the earthquake relief work since mid May; and secured power supply to the Beijing Olympic Games venues. At the same time, we kept a close eye on current economic operation status, realized well-coordinated management, and made sound judgment on the development trend.

In 2008, southern power grid maintained safe and stable operation, major production indexes were better than that of 2007, and the company accomplished ten hard-won achievements. We restored full operation from the affect of ice disaster and further strengthened grid infrastructure. We secured power supply to the Beijing Olympic Games venues and enhanced supply reliability for significant events and activities. We steadily facilitated the establishment of safety operation system and improved management and control of large-scale gird. We actively coordinated upper-stream and lower-stream industries to balance sophisticated supply-demand situation. We finished infrastructure projects on schedule and optimized grid expansion planning. We have been practicing efficient operation so as to tap new sources of income and reduce expenditures. We executed loss reduction measures in generation, transmission, distribution and consumption. We promoted energy saving and emission reduction. We reinforced internal management and improved corporate governance. Our personnel assessment was exercised in a comprehensive and detailed manner. Our sound human resources management helped to boost the entire team performance. We stressed party affairs management with innovation and promoted the construction of ideological civilization.

The core of our target in 2009 is to strive for a 5% increase in power sales to make our due contribution to the central government's 8% economic growth pledge. The following eight tasks will be our priorities:

(1) to better our understanding of the market, to increase sales and enlarge supply capacity.

(2) to increase investment and to speed up grid development in an all-round way.

(3) to improve efficiency by actively exploiting potentiality and to maintain sound operation.

(4) to grasp new characteristics of power system so as to ensure grid security and stability.

(5) to further energy saving and emission reduction, and play a supporting role in regional industrial restructuring.

(6) to continue improving internal management.

(7) to emphasize employee development and training.

(8) to improve Labors' Union with an aim to inspire employee's vitality and creativity.

The current financial crisis provides us with best learning opportunity. In the face of the complex social and economic situation, we will learn to implement scientific development theory and keep the company's progress with the time; we should seize the opportunity to strengthen and optimize southern power grid; we should act in accordance with people-oriented principle, lead our team with discipline and love; we should follow the fundamental requirement of overall coordination and sustainability and practice internal strengthening; we should progress with comprehensive planning and thorough consideration to create a harmonious environment for better development.

原文翻译

第三节 中国南方电网

根据国务院公布的"电力体制改革方案",中国南方电网有限责任公司(以下简称为 CSG)于 2002 年 12 月 29 日正式挂牌成立并开始运作。公司经营范围为广东、广西、云南、贵州和海南五省(区),负责投资、建设和经营管理南方区域电网。南方电网供电面积 100 万 km^2,供电总人口 2.3 亿人。

2009 年,公司完成售电量 5239 亿 kWh,比上年度增长 6.2%。西电东送电量 1156 亿 kWh,年增长 9.5%。公司营业收入 3136 亿元,增长 9.8%。公司耗费 1051 亿元投资固定资产。2009 年年底,公司总资产达到 4425 亿元。2009 年财富世界 500 强排名第 185 名。

在公司服务区域内的全网发电总装机容量为 1.6 亿 kW,220kV 及以上输电线路总长 7.6688 万 km。南方电网远距离、大容量、超高压输电,交直流混合运行,是全世界在运结构最为复杂、科技含量的电网之一。2009 年 6 月,海南互联工程的完工结束了海南电网以往的"孤立"局面。海南电网通过三个 32km 的海底电缆与南方电网连成整体。该海底电缆的长度位居亚洲第一、世界第二,同时该项目是中国第一个长距离、大容量 500kV 两岸跨海峡互联工程。2009 年 12 月,南方电网±800kV 直流输电工程单极顺利投产。作为世界首个±800kV 直流输电工程和中国超高压直流示范工程,超过 60%的工程设备由中国制造,单极输电量达到 2600MW。目前,南方电网已经形成"五条直流、八条交流"13 条超高压西电东送通道,最大输电能力超过 2300 万 kW。

南方电网与越南、老挝和柬埔寨接壤,与港澳紧密相连,在与相邻区域及国家发展电力合作方面具有独特的区位优势。公司作为大湄公河次区域电力合作中方执行单位，公司承诺为实现大湄公河次区域的经济发展和社会稳定提供优质的服务。南方电网经七条相互连接的线路（三条220kV，四条110kV）向越南供电，2009年送电量达41.4亿kWh，增长26.6%。2009年年底，115kV勐腊县（中国云南）到纳莫诺（老挝）的互连线路投入运行，公司开始向老挝北部的琅南塔和乌多姆赛提供电能。此外，南网与香港电网通过四条400kV线路和多条132kV线路相连，与澳门则由三条220kV线路和四条110kV线路相连。

董事长致辞：

2008年是不平凡的一年。公司经历并克服了一系列艰巨的挑战与考验，始终将国家和人民的利益当作头等大事，坚决履行中央国有企业应承担的责任。我们在雪灾时全面及时地抢修恢复电网运行；我们五月中旬投身于抗震救灾工作；我们保证北京奥林匹克运动会场馆的安全供电。与此同时，我们密切关注经济现状，实现协调管理，并对发展趋势做出正确判断。

2008年，南方电网维持安全稳定运行，主要生产指标优于2007年，并取得了十个来之不易的成就。我们抵抗雪灾影响，全面恢复运行，并进一步加强电网基础建设。我们保证奥运场馆的电能供应，提高重大事件和活动的供电可靠性。公司稳步促进安全操作系统的建立，改进大规模电网的监管和控制。我们积极协调上游和下游产业，从而平衡复杂的供需情况。公司如期完成了基础设施项目和优化电网扩展规划。我们培养了高效的运行方式，以此挖掘新的收入来源，寻求减少财政支出的办法。我们降低了发电、输电、配电、用电过程中的电能损耗。我们提倡节能减排。我们加强内部管理，改善公司治理。我们的人事评估全面而细致。我们良好的人力资源管理有助于提高整个团队的表现。我们强调党务管理的创新，促进精神文明建设。

2009年我们核心的目标是争取电力销售增加5%，为中央政府8%的经济增长率做出应有贡献。以下八个任务将是我们的重点：

（1）更全面了解市场，增加销售额，扩大供应力。
（2）增加投资，全面加快电网发展。
（3）充分利用潜力，提高效率，维持稳健运行。
（4）掌握电力系统的新特点，确保电网安全与稳定。
（5）进一步节能减排，对地区工业结构改革发挥辅助作用。
（6）深化改善内部管理。
（7）重视员工发展和培训。
（8）改善劳动工会，激发员工活力和创造力。

当前的金融危机为我们提供了良好的学习机会。面对复杂的社会及经济形势，我们将学习贯彻科学发展观，确保公司的发展进步；我们应抓住机遇，加强和优化南方电网；我们应依照以人为本的原则，有纪律与爱意地领导团队；我们应遵循全面协调、可持续发展和加强内部团结的基本要求；我们应在综合规划、全面考虑的基础上发展进步，为更好地发展创造和谐的环境。

New Words and Expressions

China Southern Power Grid　　　　　　　中国南方电网

west-east transmission	西电东送
installed generation capacity	装机容量
submarine cables	海底电缆
grid infrastructure	电网基础设施

Notes

China Southern Power Grid Co., Ltd. (hereinafter referred to as CSG) was established on December 29th, 2002 in accordance with "The Power Sector De-regulatory Reform Program" promulgated by the State Council of China. CSG invests, constructs and operates power networks in Guangdong, Guangxi, Yunnan, Guizhou and Hainan provinces and regions.

根据国务院公布的"电力体制改革方案",中国南方电网有限责任公司(以下简称为CSG)于2002年12月29日正式挂牌成立并开始运作。公司经营范围为广东、广西、云南、贵州和海南五省(区),负责投资、建设和经营管理南方区域电网。

Exercises

Ⅰ. Choose the best answer into the blank.

1. Which is not included in the operating range of the Southern Power Grid ?____
   A. Guangdong   B. Guangxi   C. Guizhou   D. Jiangxi

2. The project is the first long-distance, large-volume 500kV cross-strait interconnection project was built by____.
   A. Southern Power Grid   B. State Grid
   C. The State Council   D. Hainan

3. As the world's first ±800kV DC transmission project and China's UHV DC demonstration project was completed in ____.
   A. 2007   B. 2008   C. 2009   D.2010

4. In 2009 the central goal of the southern power grid was to gain an increase of ____ in electricity sales.
   A. 8%   B. 5%   C. 6.2%   D. 9.8%

5. Which of the following is not one of the eight tasks?____
   A. to better our understanding of the market, to increase sales and enlarge supply capacity.
   B. to improve the staff's amateur life by holding more activities
   C. to improve Labors' Union with an aim to inspire employee's vitality and creativity.
   D. to continue improving internal management.

Ⅱ. Answer the following questions according to the text.

1. Please give a brief introduction to the Southern Power Grid.

2. Compared with 2008, what progress in the Southern Power Grid in 2009.

3. Please describe briefly the general situation and significance of the Hainan Interconnection Project.

4. Why is it said that the southern power grid has geographical advantages?

III. Translate the following into Chinese.

Based on the analysis on historical data of power supply reliability of main cities supplied by China Southern Power Grid and considering the development stage of these urban power networks as well as the economic development of these cities, by means of comprehensively applying both historical data based power supply reliability trend forecasting method and power supply reliability forecasting method that takes relevant impacting factors into account, the annual power supply reliability indices of these cities during the 11th Five-Year Plan are forecasted; according to the principle of relevance and combining with concrete circumstances of urban power networks, the reasonable target values of power supply reliability are determined. To lead rational investment in urban power supply enterprises and improve power supply reliability level of main cities supplied by China Southern Power Grid, the technical and management measures to implement power supply reliability objectives are proposed.

Keys

I. Choose the best answer into the blank.

1. D   2. A   3. C   4. B   5. B

II. Answer the following questions according to the text.

1. China Southern Power Grid was established on December 29th, 2002. CSG invests, constructs and operates power networks in Guangdong, Guangxi, Yunnan, Guizhou and Hainan provinces and regions. The service area is of 1 million square kilometers, with a population of 230million.

2. In 2009, CSG's electricity sale was 523.9TWh, a 6.2% increase than the previous year. The west-east transmission reached 115.6TWh, recording a 9.5% increase. The company revenues reached 313.6 billion RMB, increased by 9.8%.

3. the completion of Hainan Interconnection Project ended the long history of "isolated" Hainan grid. Hainan grid was integrated into southern power grid by three 32km submarine cables. The submarine cables are the longest in Asia and the second longest in the world. The project is the first long-distance, large-volume 500kV cross-strait interconnection project in China.

4. Because being adjacent to Hong Kong, Macau, and bordering Vietnam, the Laos and Cambodia, CSG service area enjoys unique geographical advantage in developing power cooperation with neighboring regions and countries.

Ⅲ. Translate the following into Chinese.

在分析南方电网各主要城市供电可靠性历史数据的基础上，结合各城市电网所处的发展阶段以及城市经济发展情况，综合应用基于历史数据的供电可靠性趋势预测方法和考虑相关影响因素的供电可靠性预测方法，"十一五"期间南方电网各主要城市年度供电可靠性指标进行了预测，根据相关原则并结合电网的具体情况综合考虑确定合理的供电可靠性目标值。提出了实现供电可靠性目标的技术措施和管理措施，旨在引导城市供电企业的合理投资，提升南方电网各主要城市的供电可靠性水平。

翻译技巧之句型转换

在翻译的过程中，可以根据原文的具体情况，按照汉语的表达习惯，对原文句子结构进行调整和转换。下面介绍其他几种较常用的转换方式：

1. 否定变肯定译法

该方法是将原文中否定的表达形式译成肯定形式。例如：

Until recently geneticists were not interested in particular genes. 基因学家们最近才开始对特定基因感兴趣。

Don't start working before having checked the instrument thoroughly. 要对仪器彻底检查才能开始工作。

The flowing of electricity through a wire is not unlike that of water through a pipe. 电流过导线就像水流过管子一样。

2. 肯定变否定译法

是将原文中肯定的表达形式译成否定形式。例如：

The influence of temperature on the conductivity of metals is slight. 温度对金属的导电性影响不大。

As rubber prevents electricity from passing through it, it is used as insulating material. 由于橡胶不导电，所以用作绝缘材料。

There are many other energy sources in store. 还有多种其他能源尚未开发。

3. 主动变被动译法

在具体文章中，如果原文主动句正面译出比较困难，或者译成汉语被动句更能准确表达原意、汉语行文更为方便时，可对原文主动句译成汉语的被动句。

The properties of materials have dictated nearly every design and every useful application that the engineer could devise. 工程师所能设想的每一种设计和每一种用途几乎都要受到材料性能的限制。

Since prehistoric times the sketch has served as one of man's most effective communication techniques. 自从史前时期以来，草图一直被人类用作最有效的交际手段之一。

References

[1] COLLINS M M C. Electric-power Transmission [M]. Canada: The Canadian Encyclopedia, 2006.

[2] SUN T, XIA J L, SUN Y Z, et al. Research on the applicable range of AC and DC transmission voltage class sequence[C].International Conference on Power System Technology, 2014, 374-380.

[3] ZHOU X, YI J, SONG R, et al. Haiyuan Tang. An overview of power transmission systems in China [J]. Energy, 2010, 35(11):4302-4312.

[4] YANG F L, YANG J B, ZHANG Z F. Unbalanced tension analysis for UHV transmission towers in heavy icing areas [J], Cold Regions Science & Technology, 2012, 70(1):132-140.

[5] MENG X B, WANG L M, HOU L, et al. Dynamic characteristic of ice-shedding on UHV overhead transmission lines[J],Cold Regions Science & Technology, 2011, 66(1)44-52.

[6] JIA J J,WANG H CHEN X D, et al. Neural Network Simulating the Tensility of Transmission Line Ice Shedding[J],Energy Procedia, 2012, 17(1):1235-1241.

[7] CHEN L, MACALPINE J M K, BIAN X, et al.Comparison of methods for determining corona inception voltages of transmission line conductors [J]. Journal of Electrostatics, 2013, 71(3):269-275.

[8] TANIGUCHI S, OKABE S, TAKAHASHI T, et al. Air-Gap Discharge Characteristics in Foggy Conditions Relevant to Lightning Shielding of Transmission Lines [J]. IEEE Transactions on Power Delivery, 2008, 23(4):2409-2416.

[9] ROBERT D, C. Overview of the transmission line design process [J]. Transmission Systems Engineering and Research Design and Construction Division, Los Angeles Department of Water and Power, 1995, 3:109-118.

[10] VAJJHALA S P, FISCHBECK P S. Quantifying siting difficulty: A case study of US transmission line siting [J]. Energy Policy, 2007(35): 650-671.

[11] FENTON G A, SUTHERLAND N. Reliability-Based Transmission Line Design [J]. IEEE TRANSACTIONS ON POWER DELIVERY,2011(26):596–606.

[12] AGGARWAL R K, JOHNS A T, JAYASINGHE J A S B, et al. An overview of the condition monitoring of overhead lines [J]. Electric Power Systems Research, 2000, 53(1):15-22.

[13] HAMZA A S H A, ABDELGAWAD N M K, ARAFA B A. Effect of desert environmental conditions on the flashover voltage of insulators [J]. Energy Conversion & Management, 2002, 43(17):2437-2442.

[14] TENA G M, RAMIRO H C, JORGE I M T. Failures in outdoor insulation caused by bird excrement [J]. Electric Power Systems Research, 2010, 80(6):716-722.

[15] CHRISTODOULOU C A, EKONOMOU L, FOTIS G P, et al.Optimization of Hellenic overhead high-voltage transmission lines lightning protection [J]. Energy, 2009, 34(4):502-509.

[16] CHANAKA M, SHANTHI K, PERERA R. Modeling of Power Transmission Lines for Lightning Back Flashover Analysis [C] //International Conference on Industrial and Information Systems, 2011:386-391.

[17] ZHOU X, YI J, SONG R, YANG X, et al. An overview of power transmission systems in China [J] .Energy, 2010, 35(11):4302-4312.

[18] HLEDIK R. How Green Is the Smart Grid [J]. Electricity Journal, 2009, 22(3):29-41.

输电线路工程系列教材

书名	作者
输电线路工程概论	祝 贺
输电线路基础	李光辉
架空输电线路设计（第二版）	孟遂民
输电杆塔及基础设计（第三版）	陈祥和
输电铁塔设计	安利强
高压架空输电线路施工（第二版）	祝 贺
输电线路施工机械	葛永庆
输电线路运行维护理论与技术（第二版）	白俊锋
输电线路试验理论与技术	陈井彦
架空输电线路运行与检修	罗朝祥
输电线路施工与运行维护	黄宵宁
输电线路金具理论与应用	赵 强
电力线路金具基础与应用	李光辉
电力电缆施工技术（第二版）	李光辉
配电线路设计施工、运行与维护（第二版）	李光辉
电网建设工程造价控制与管理（第二版）	葛 乐
输电线路工程概预算	江全才
输电线路地理信息系统	张广洲
输电线路 CAD	苑素玲
输电线路电磁环境	张广洲
工程有限元基础	安利强
电力工程测量技术	李 利
输电线路工程专业英语	唐 波
输电线路工程课程设计	祝 贺
输电线路工程综合实验	江文强

中国电力出版社官方微信

中国电力教材服务官方微信

◀ 请关注中国电力教材服务官方微信，获取更多教材资源

中国电力出版社教材中心
教材网址　http://jc.cepp.sgcc.com.cn
服务热线　010-63412548　63412523

ISBN 978-7-5198-0750-4

定价：33.00 元